Aeronautics - Characteristics and Emerging Technologies

Edited by Longbiao Li

Published in London, United Kingdom

Aeronautics - Characteristics and Emerging Technologies
http://dx.doi.org/10.5772/intechopen.1003414
Edited by Longbiao Li

Contributors
Abhay Dhasmana, Anatolii Kretov, Anoy Saha, Dmytro Tiniakov, Drago Sever, Edgar Jeevan Danaraj, Goran Vorotović, Jela Vorotović, Leila Halawi, Longbiao Li, Mark Miller, Mona Ghassemi, Nebojša Petrović, Nikola Davidović, Pavle Petrović, Sachin Srivastava, Sam Holley, Sergej Težak, Đorđe Novković

First published in London, United Kingdom, 2025 by IntechOpen
IntechOpen is the global imprint of INTECHOPEN LIMITED, registered in England and Wales, registration number: 11086078, 167-169 Great Portland Street, London, W1W 5PF, United Kingdom

For EU product safety concerns: IN TECH d.o.o., Prolaz Marije Krucifikse Kozulić 3, 51000 Rijeka, Croatia, info@intechopen.com or visit our website at intechopen.com.

British Library Cataloguing-in-Publication Data
A catalogue record for this book is available from the British Library

Aeronautics - Characteristics and Emerging Technologies
Edited by Longbiao Li
p. cm.
Print ISBN 978-0-85014-859-6
Online ISBN 978-0-85014-858-9
eBook (PDF) ISBN 978-0-85014-860-2

Meet the editor

Dr. Longbiao Li is a lecturer in the College of Civil Aviation at the Nanjing University of Aeronautics and Astronautics in China. Dr. Li's research focuses on the vibration, fatigue, damage, fracture, reliability, safety and durability of aircraft and aero engines. In this research area, he has published 200+ SCI journal publications, 12 monographs, 4 edited books, 4 textbooks, 3 book chapters, 34 Chinese patents, 2 US patents, 6 Chinese software Copyright, and more than 30 refereed conference proceedings. He has been involved in different projects related to structural damage, reliability, and airworthiness design for aircraft and aero engines, supported by the Natural Science Foundation of China, COMAC Company, and AECC Commercial Aircraft Engine Company.

Contents

Preface

Aeronautics, the science and technology of flight, has witnessed remarkable advancements in recent years. This field encompasses a wide range of characteristics and emerging technologies that are shaping the future of aviation. One of the key characteristics of aeronautics is the pursuit of enhanced performance. Aircraft are designed to be more fuel-efficient, reducing the environmental impact and operating costs. Advanced materials, especially composites, and aerodynamic designs contribute to improved lift and reduced drag, enabling aircraft to fly more efficiently and at higher speeds. Another important aspect is safety. The aviation industry places a premium on ensuring the safety of passengers and crew. This involves continuous improvements in aircraft systems, including advanced avionics, enhanced structural integrity, and reliable propulsion systems. Stringent safety regulations and rigorous testing procedures also play a crucial role in maintaining a high level of safety in the aerospace sector. Emerging technologies are driving significant changes in aeronautics. One such technology is the development of electric and hybrid-electric propulsion systems. These systems offer the potential for reduced emissions and quieter operations, making aviation more sustainable.

Additionally, artificial intelligence and machine learning advancements enable more efficient flight planning, maintenance scheduling, and air traffic management. In conclusion, aeronautics is a dynamic field that combines traditional characteristics with cutting-edge technologies. The future of aeronautics is bright, with safety and new technologies leading the way.

Longbiao Li
College of Civil Aviation,
Nanjing University of Aeronautics and Astronautics,
Nanjing, PR China

Section 1

New Trend in Aerospace Engineering

Chapter 1

UAVs and Hydrogen Technologies are the Future of Civil Aviation

Longbiao Li, Anatolii Kretov and Dmytro Tiniakov

Abstract

Nowadays there are a lot of startups about creation of UAVs for civil application. Most of them are based on electric (batteries) or traditional fuel (oil-based) technologies. From other side, requirements of ICAO about reducing environmental impact of civil aviation are pushing aviation companies to develop green technologies. Some of them are based on hydrogen fuel application. Application of hydrogen can reduce the environmental impact of aviation. The aim of this chapter is to analyze possible trends of civil unmanned aircraft application and to evaluate the possibility and efficiency of hydrogen fuel application for such drones. It was demonstrated that the implementation of hydrogen technology in civil drones presents significant challenges that necessitate comprehensive examination. The two technologies, when considered separately, offer a plethora of advantages. However, when combined, they can potentially lead to undesirable outcomes. On the one hand, they can contribute to an improvement in the environmental impact. On the other hand, they possess unique characteristics that interact with one another. It has been demonstrated that the integration of hydrogen fuel can result in an 8% increase in the mass of the drone, which subsequently reduces its overall efficiency.

Keywords: aircraft, UAV, hydrogen technology, takeoff mass, civil aviation

1. Introduction

Currently, civil aviation is at the peak of its development. Thousands of aircraft carry millions of passengers and hundreds of tons of cargo every day (see **Figure 1**) [1]. It would seem that there is nothing that could be improved in modern commercial aviation. But there are several challenges ahead.

The first is related to a problem common to all humanity: environmental impact. Of course, aviation does not produce much pollution compared to other industries or modes of transportation (see **Figure 2**) [2]. However, reducing even a relatively small amount of emissions can improve environmental protection. ICAO has developed a set of incentives for civil aviation that should ensure a reduction of about 50% in pollution and noise by 2050. There are several approaches to meeting the ICAO requirements (see **Figure 3**) [3]. One is the use of fuel based on hydrogen technology. This solution can significantly reduce the environmental impact of aviation. However, hydrogen technologies face several difficulties in their application to civil aviation [4].

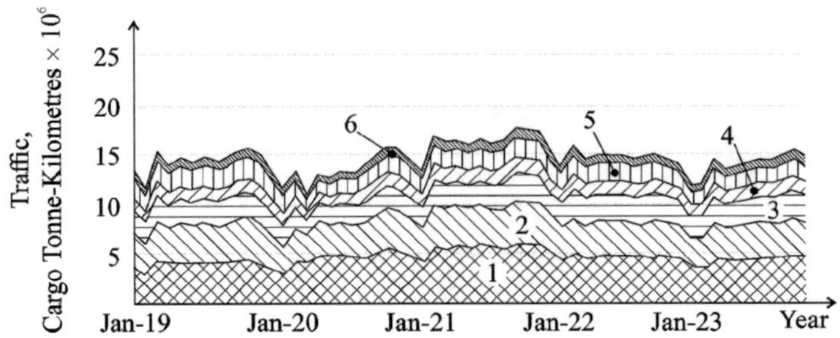

Figure 1.
Cargo traffic [1]: 1—Far East to/from North America; 2—Europe to/from Far East; 3—Europe to/from North America; 4—Middle East to/from Far East; 5—Intra Far East; 6—Europe to/from Middle East.

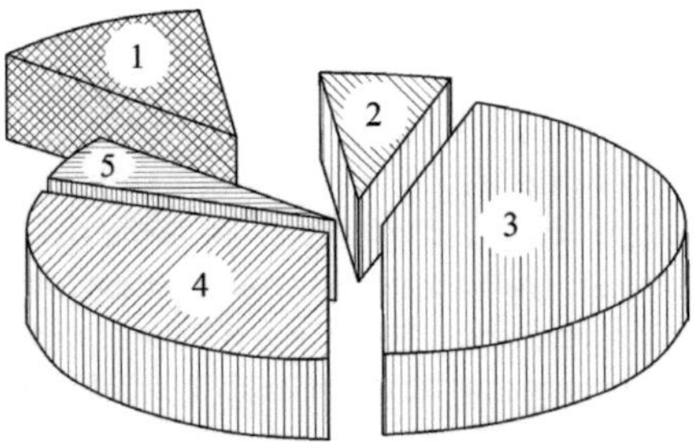

Figure 2.
Amount of CO_2 emissions distributed among different transportation categories [2]: 1—Aviation 12%; 2—Maritime 7%; 3—Light-duty road vehicles 45%; 4—Heavy-duty road vehicles 30%; 5—Rail, Inland water 6%.

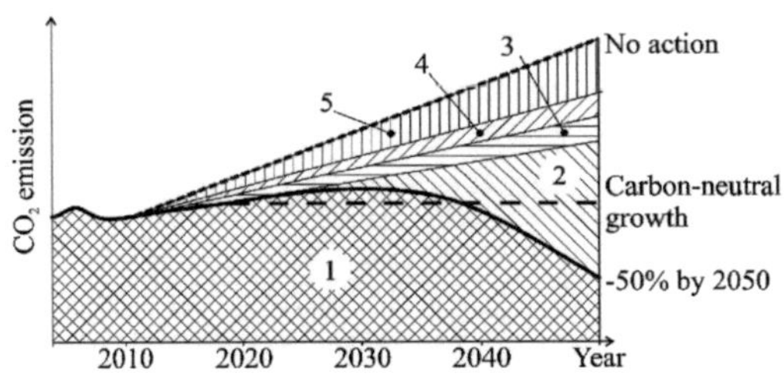

Figure 3.
Aviation industry roadmap for emissions mitigation [3]: 1—Net emission trajectory; 2—Additional technologies and fuel; 3—Infrastructure; 4—Operations; 5—Technology.

The main problem is to ensure the safe operation of aircraft with hydrogen-based propulsion systems.

As illustrated in **Figure 3**, the implementation of novel fuel technologies (2), engineering innovations (5), and aviation operations strategies can collectively yield the desired outcomes in reducing pollution. Let us examine the potential of these new technologies in aviation. One avenue for consideration is the optimization of existing aircraft design solutions. These researches [5, 6] exemplify cases where such approaches were contemplated, resulting in an estimated 7—9% reduction in emissions.

However, it is not possible to achieve the required reduction of pollution (50%) at this rate.

Another possible solution is the use of unmanned aviation. This approach allows for the following specifics: possibility of high level of automation, relatively low design, manufacturing and operating costs compared to traditional aircraft; etc.

Figure 3 illustrates that new technologies for fuels can provide the most significant contribution to reducing emissions. There are several approaches in this direction.

The first is related to the transition to electric power, which is widely applied for modern UAVs. However, the capacity of today's batteries is not yet sufficient to carry passengers and cargo typical of modern commercial aviation. Nevertheless, numerous initiatives have been undertaken to implement electric propulsion systems for general aviation aircraft and to short-haul aircraft [7, 8]. It is important to note that the majority of electricity is currently generated by burning fossil fuels, resulting in CO_2 emissions of 10 billion tons per year in the global power industry. In light of these considerations, it is imperative to discuss the environmental friendliness of electrically-powered UAVs with an appropriate caveat and to consider the manner in which electricity is generated.

A second approach that is also being widely explored for commercial aviation is the use of hydrogen fuel. This is not a novel concept [9–11]; however, these technologies are not yet sufficiently developed for commercial aviation and require a comprehensive analysis of potential design and production solutions, as well as improvements to ground infrastructure.

Consequently, at this juncture, it is possible to identify two of the most promising technologies that, when combined, have the potential to achieve a significantly greater effect than either technology could achieve when used separately [12]. Each of these approaches has its own particulars. The combination of the two may yield unanticipated outcomes. The objective of the proposed research is twofold: first, to elucidate the distinctive characteristics of both technologies; second, to examine the impact of hydrogen fuel on the take-off mass of a drone.

Let us now examine these technologies in greater detail.

2. UAV technologies

UAVs are a type of flying vehicle. They are used in the civil sector by commercial enterprises and individuals and are part of aviation complexes for various purposes (commercial and non-commercial). The manufacturing of UAVs belongs to the high-tech production of the aviation industry. Due to the high added value, the development of UAV production is a priority for high-tech and developed countries.

The creation and mass production of UAVs for commercial and non-commercial applications, as well as the trends in the development of UAVs worldwide, are of significant relevance and impact the aviation industry (See Section 1).

2.1 A brief history of UAVs

It is unfortunate that, as the history of aviation technology development shows, the main driving force of progress is its military application. Nevertheless, the advent of civilian aircraft is inevitable. A similar phenomenon occurred with the advent of UAVs. The advent of UAVs has led to the emergence of new possibilities in military operations. These vehicles have been employed in a variety of roles, including tactical reconnaissance, the delivery of combat weapons (such as bombs and torpedoes) to designated locations, and the management of combat operations.

The background of the use of drones is deep in history. In China, kites carrying bombs were used to attack the walled cities. In ancient Greece, the case is known when it was built a steam-powered "pigeon" that apparently could fly. The Egyptian Saqqara bird, a bird-shaped artifact, may have been able to glide [13].

The inaugural documented instance of the utilization of an unmanned aerial vehicle (UAV) occurred during the Austrian siege of Venice in 1849 [13].

The term "drone" is used to describe a device that is controlled remotely. The concept of a drone was first proposed by Nikola Tesla, who filed a patent application for a vehicle that could be controlled by radio waves [13].

The next stage in the development of drones can be considered the unmanned flying target (Fairy Queen biplane), which was developed in Britain in 1931 [14].

A further example of a drone is the German V-1 flying bomb, the success of which led not only to the development of missile technology but also autonomous aircraft [15].

For a considerable period prior to their wider application, UAVs were employed primarily for military purposes. It is worthy of note to mention the Ryan Model 147E reconnaissance drone, developed in the USA in 1963 [16], as well as drones developed in the USSR by Tupolev [17]. These drones are of a considerable size, with dimensions that are equivalent those of conventional aircraft with comparable functionality.

The use of radio-controlled drones was first developed in Israel in 1982. The most commonly utilized models were the IAI Scout and Tadiran Mastiff [17]. These drones were developed from the ground up, rather than being converted from existing aircraft.

Subsequently, other countries, including the USA, have commenced the production of similarly military drones, such as the MQ-1B Predator and MQ-9 Reaper [17].

In the latter half of the twentieth century, the development of radio-controlled drones for domestic use commenced in earnest. In the beginning, these models were primarily constructed by individuals for their own purposes, with a resemblance to aircraft. However, this sector of production then experienced a period of rapid expansion. UAVs have a multitude of applications, extending beyond their military and individual uses.

The company Boeing, engaged in the development of a passenger airliner scheme designated Blended Wing Body (BWB), recognized the necessity to create a demonstrator of this novel scheme in the form of a flying laboratory. In particular, a 1:8 scale UAV designated X-48C was constructed and operated [18].

2.2 A brief UAVs classification

Drones are UAVs that are autonomous robotic systems. Their primary function is to perform flights that are potentially dangerous for humans, according to a predetermined program. This program allows for automatic or manual adjustment of the flight task, as well as operational decision-making, in response to changing flight conditions and the surrounding space.

UAVs may exhibit varying degrees of autonomy, as well as differences in design, purpose, and other parameters.

At present, there are no industry standards for categorizing drones. Manufacturers are not constrained by any standards, and as a consequence, there are no regulatory requirements for drone equipment at this time. This may be attributed to the fact that UAVs already possess a variety of configurations, aerodynamic designs, and

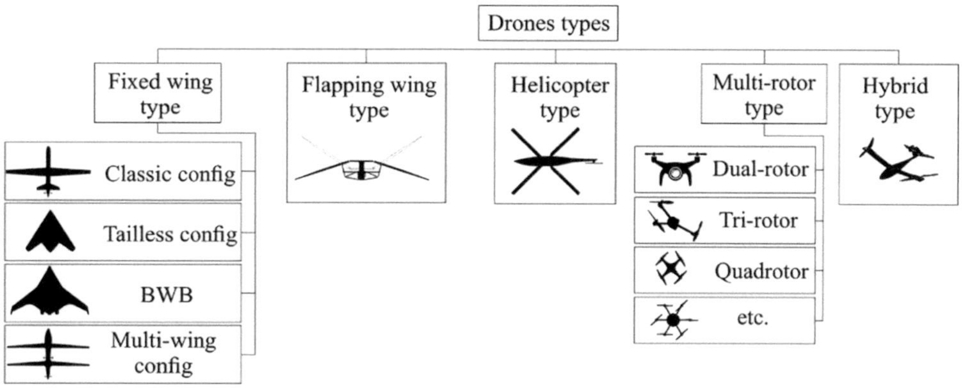

Figure 4.
Brief classification of drones by aerodynamic principle.

Type	NASA UAS Class	Mass, kg	Operating altitude, m	Mission range, km	Flight endurance, h	Payload, kg
Micro	sUAS Class I	< 2	< 140	~ 5	< 1	< 1
Mini		2—25	< 1000	~ 25	2—8	< 10
Small		25—150	< 1700	~ 50	4—12	< 50
Medium	Class II	150—600	< 3300	200—500	8—20	< 200
Large	Class III	> 600	> 3300	> 1000	> 20	> 200

Table 1.
UAV classification based on mass, altitude, range, and payload [19].

components. Consequently, the implementation of rigid standards and classifications may impede the advancement of UAV technology.

Drones can be broadly classified based on their aerodynamic principle, weight, size, range, frame type, control method, purpose, etc. (See **Figure 4**, **Table 1**).

It is important to note that this type of aircraft is undergoing rapid development, with models that have already surpassed traditional classification. For instance, the RQ-4 Global Hawk [17], with a mass of over 11 tons, represents a significant advancement in this field. Additionally, there are heavier models currently in development.

2.3 UAV engine power sources

The drones utilize energy sources that are well-known in the field of aviation. They are traditional aircraft fuel, batteries, fuel cells, and solar cells.

The majority of aircraft fuel (kerosene or petroleum) is utilized in the operation of large drones. The aforementioned drones necessitate considerable flight endurance, coupled with a relatively high payload capacity. In the contemporary time, only traditional aviation fuel can provide the requisite performance. A number of studies have been conducted with the objective of implementing electrical batteries in such UAVs. However, these studies are still in the developmental stage [20].

Batteries are utilized primarily in the operation of smaller UAVs. These drones are designed for short-range operations with relatively low payloads. They require less operational time than drones that utilize traditional aviation fuel. The use of rechargeable or replaceable batteries allows for a reduction in maintenance time [20].

A fuel cell is a type of electrochemical unit. It is converting chemical energy from a chemical substance directly into electrical energy. The conversion of thermal to mechanical energy is absent, and this energy source has high efficiency and is environmentally preferable. Fuel cells are currently not commonly utilized in UAVs. Due to the relatively high weight of the cell, large fixed-wing or rotor UAVs are the only types that can be equipped with such a cell. A significant advantage of fuel cells in UAVs is the extended operational range that they enable, obviating the need for frequent recharging [20].

The use of solar cells in drones is a relatively uncommon phenomenon within the current drone industry. The majority of drones that employ solar cells are wing drones. The drones in question have sufficient surface area for the solar panels to be arranged on. In consideration of the low efficiency of current solar panels, they are typically best suited for multi-motor drones. It should be noted, however, that solar cells are suitable for small flapping-wing drones [20].

2.4 UAVs application and payload types

The majority of payloads can be attached to or arranged within UAVs, with the only restrictions typically being the power of the UAV's motors and the size of the payloads. The majority of drones are equipped with cameras by their manufacturers. Nevertheless, other equipment can be procured from drone manufacturers, and drone users are also able to affix or arrange payloads into their drones independently [21].

It is first necessary to consider the various kinds of sensors that can be used as payloads for UAVs. Cameras and microphones are nowadays standard equipment, but drones can also be equipped with other sensors, such as night vision cameras, infrared cameras, chemical and biological sensors, and meteorological sensors (measuring wind, temperature, humidity, and pressure). All of the above devices allow UAVs to be widely used to monitor various conditions, spaces, and objects. Some examples of the types of monitoring that can be carried out are given below.

The use of UAVs for public order purposes is becoming increasingly common [21]. These vehicles can be used to monitor road traffic, guarded objects, and other external conditions. They can also be employed to monitor weather and the environment, including the measurement of air pollution, fires, floods, and animal migration. Furthermore, UAVs can be used to monitor the condition of far, long, and hard-to-access infrastructure objects, such as power lines, pipelines, bridges, and so on.

Secondly, a variety of equipment can be mounted on drones, including cargo holders, manipulators, and specialized equipment [21].

One illustrative example is the use of drones for the delivery of cargo. This service is already in use in some countries and certain regions. In the field of agriculture, drones can be utilized for a multitude of purposes, including the monitoring of plant and crop conditions, the application of fertilizers, the spraying of chemicals against pests, and even the harvesting of crops. As an illustration of infrastructure maintenance, we may consider the maintenance of power lines, specifically the cleaning of wires and insulators from dirt. There have been developments in the use of drones for the urgent delivery of medical equipment to remote and hard-to-reach areas. The efficacy of drones in extinguishing fires has already been demonstrated. To maintain

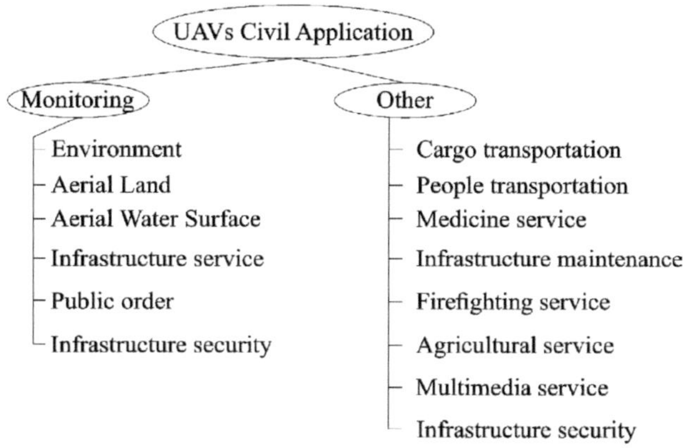

Figure 5.
Brief list of where modern UAVs get used.

public order, drones can be equipped with stun guns that can be deployed remotely at the discretion of the drone operator.

Furthermore, the utilization of UAVs for transportation is already a reality. It is widely acknowledged that road traffic accidents are a significant concern, particularly in comparison with the relatively low accident rate observed in aviation [22]. However, if we replace the human driver with a ground operator (who has undergone extensive training) or artificial intelligence, in other words, with an autopilot, this problem (high accident rate) can be avoided.

Consequently, the application of UAVs represents a promising technological field (**Figure 5**), particularly in the event that the advancement of electric battery technology (or other green fuels) and the development of autopilot technology are achieved.

3. Hydrogen technologies

As previously stated (See Section 2), hydrogen can be employed as an energy source in both heat engines and fuel cells, with the objective of powering electric motors. The most notable application of hydrogen fuel cells was demonstrated on the Apollo [9–11] spacecraft during the lunar space program. In many respects, modern fuel cells are superior to lithium-ion batteries. However, they are still quite difficult to manufacture, which is reflected in their high cost. For this reason, the authors posit that the most promising avenue for the use of hydrogen in UAVs is its incorporation into thermal engines.

The potential for hydrogen fuel to be used in aircraft engines in the future is due to the fact that hydrogen is a renewable energy source with the following main thermodynamic and specific properties:

- Hydrogen has a significant calorific value, with a specific heat of combustion that is three times higher than that of kerosene. This is evidenced by the fact that hydrogen has a specific heat of combustion of 120.9—106 J/kg, while kerosene has a specific heat of combustion of 40.8—106 J/kg [23].

- Furthermore, liquid hydrogen has the potential to effectively cool hot engine parts, due to its high cooling resource. This is advantageous in situations where engine parts have been exposed to high temperatures.

- The combustion process is characterized by stability in the presence of poor fuel mixtures.
- It has a high combustion rate.

For instance, it provides a jet engine with an exceptionally high specific thrust-to-impulse ratio, which is to say, an optimal relationship between engine thrust and fuel consumption. This is a direct consequence of the record energy intensity and low mass of hydrogen. In the combustion chamber, hydrogen burns steadily and does not create any harmful pulsations.

Moreover, hydrogen environmental friendliness is its most significant advantage. The exhaust is primarily water vapor with a small admixture of nitrogen oxides [24].

3.1 An overview of hydrogen in the aerospace industry

The concept of utilizing a fuel blend of liquid hydrogen and liquid oxygen was initially proposed by Tsiolkovsky at the dawn of the twentieth century [25]. This concept was not seriously contemplated until after the conclusion of World War II.

In 1957, the United States successfully tested the J-57 Pratt & Whitney engine. In the same year, a B-57 aircraft with a hydrogen fuel system was tested at the Lewis Research Center (USA) and demonstrated successful flight performance on multiple occasions [26].

In November 1963, the United States successfully launched the Atlas-Centaur launch vehicle, which was the first in the world to utilize an oxygen–hydrogen upper stage [27].

Subsequently, the Space Shuttle employed a similar configuration of thrusters.

In the Soviet Union, the RD-0120 oxygen–hydrogen engine, developed for the Energia launch vehicle, achieved the greatest degree of notoriety [28].

Another application of this technology was achieved in the USSR: the cryogenically fueled airplane, the Tu-155 (**Figure 6**) [11, 29]. The aircraft took off in 1988, utilizing a combination of conventional and cryogenic propulsion systems. One of the three engines, located on the right, was replaced by a hydrogen NK-88. The development of the Tu-155 was closely related to the development of the Energia-Buran program, which also created a hydrogen engine, albeit a rocket engine [10].

The project was continued in 1998 with the commencement of tests on the Tu-156 aircraft using liquefied natural gas (LNG) as fuel. According to the estimates of Lockheed and the results of flight tests conducted by researchers at the Tupolev Company on the Tu-156 aircraft, hydrogen fuel-powered engines will have a resource of approximately 25% more than kerosene-powered engines due to the absence of soot formation in the gas-air path [23].

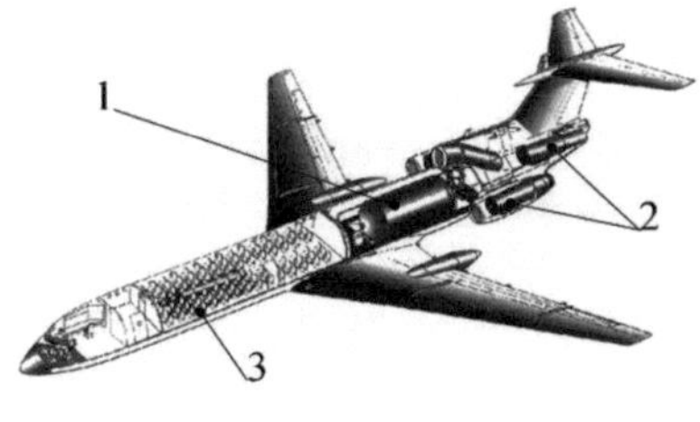

Figure 6.
Tu-155 is the inaugural transport category aircraft to be equipped with a hydrogen fuel power unit [10]: 1—Hydrogen fuel tank; 2—Engines; 3—Passenger compartment.

DOI: http://dx.doi.org/10.5772/intechopen.1008352

A concept study on a hydrogen fuel cell blended-wing-body aircraft was conducted by NASA between 1999 and 2004. The basic design requirements used for the study included a payload capacity of 225 passengers (with a three-class seating arrangement) plus baggage and a range of 6500 km with a full payload [30].

In 2005, the Naval Research Laboratory initiated research on the Ion Tiger UAV project [31]. The aircraft is powered by hydrogen cells. The aircraft completed approximately 23 hours of flight time in 2009, with a take-off mass of 16 kilograms and a payload of 2 kilograms.

In 2022, the Australian company Hypersonix Launch Systems proposed a concept design of a hypersonic satellite delivery aircraft to orbit [32]. The Delta Velos vehicle is equipped with four Spartan hypersonic direct-flow air-jet engines at once. In this design, hydrogen does not require a supply of oxygen on board for combustion; the necessary oxidizer is extracted from the air stream at speeds greater than M = 1. A specially designed Spartan engine duct compresses the rushing air stream to such an extent that the injected hydrogen in the combustion chamber is immediately ignited. The first flight for the test model is planned for the end of 2024 or the first quarter (Q1) of 2025.

However, according to experts in the field of cryogenic fuels, the widespread use of hydrogen as an alternative to kerosene in civil aviation in developed countries is not expected until 2030.

3.2 Hydrogen technology key specifics

The utilization of hydrogen fuel necessitates the implementation of specific requirements pertaining to the aircraft layout. Liquid hydrogen has a lower density than traditional fuel, such as kerosene. The density of liquid hydrogen is 0.07 g/cm^3, while that of kerosene is 0.82 g/cm^3 [33]. Consequently, to achieve the same thermal equivalence, it is necessary to increase the capacity of fuel tanks for fuel storage by approximately 3.8 times, while reducing the total mass of fuel by approximately 3 times. It is also necessary to ensure the reliable thermal insulation of fuel tanks to prevent the boiling out of liquid hydrogen, which could result in the rupture of the tanks.

In the event of a leak from the line, hydrogen has the potential to form an explosive mixture with oxygen. The explosion limits of a hydrogen-oxygen mixture are in the range of 4–94% H2 by volume, while for a hydrogen-air mixture, the range is 4–74% H2 [33]. The mixture of two volumes of H_2 and one volume of O_2 was designated explosive gas due to its exceptional explosiveness. Consequently, the utilization of hydrogen fuel necessitates, among other factors, the attainment of higher qualifications among operating personnel and the implementation of adjustments to ground infrastructure.

The aforementioned favorable attributes and the fact that hydrogen is a renewable energy source render it an optimal fuel for the future. However, the implementation of hydrogen fuel in aviation is contingent upon further research and the enhancement of both aircraft and ground infrastructure.

4. A concept analysis of the potential for utilizing cryogenic fuel in UAVs

4.1 Introduction

As demonstrated in Sections 2 and 3, UAVs and hydrogen technologies are not novel in the engineering field. Based on the classification of UAVs presented in Section 2 and the characteristics of liquid hydrogen outlined in Section 3, it can be inferred that liquid hydrogen fuel is not suitable for small UAVs. This technology necessitates specific conditions regarding temperature, additional volume for storage, and other factors. Therefore, it is essential to consider these particulars in numerical research.

The storage of hydrogen fuel necessitates the maintenance of specific low-temperature conditions, reaching as low as −253°C. In certain projects (see Section 3), LNG was utilized. It is capable of withstanding temperatures of −162°C [33], which are more conducive to its storage. Such fuels are collectively referred to as cryogenic fuels (CF). A comparative analysis of the traditional fuel (kerosene) and CF (liquid hydrogen and LNG) would be a reasonable course of action. Such an analysis can facilitate a more nuanced understanding of the impact of storage temperature requirements on the deployment of novel fuels.

Whether it is justifiable to use CF in UAVs is a fundamental one that should be addressed at the outset. Given the limited initial data available to the designer at the outset, the methodology for justifying the use of CF for UAVs should be clear and reliable and capable of withstanding scrutiny. Existing studies in this field [33–37] commonly employ reliable software products, which typically necessitate extensive preliminary research and a meticulous examination of potential variants.

The prevailing opinion among experts is that the use of CF in drones should be limited to fuel cells, specifically electric traction. In consideration of the collective experience gained in the creation of a flying laboratory based on the Tupolev Tu-155 aircraft with engines operating on CF (See Section 3), the authors sought to conduct a mass evaluation of the utilization of CF for UAVs equipped with traditional aviation engines. Many authoritative experts have identified the use of hydrogen as a fuel as the primary factor in developing a comprehensive strategy for future aviation, including unmanned aircraft. Nevertheless, it is challenging to achieve such a transition in a single step. In light of this, it is advisable to contemplate an intermediate stage, which is indispensable for the analysis and implementation of expertise in the design and operation of carbon fiber for the UAV with CF. This is particularly relevant in the context of the initial use of LNG.

As illustrated in **Table 2**, the principal attributes of the fuel required for a preliminary evaluation against traditional fuel for gas turbine engines are presented. The aviation kerosene Jet A-1 serves as a benchmark for comparison.

The Tupolev company's application of CF in the 1980s constituted a seminal effort in its utilization, as detailed in Section 3. Nevertheless, the developers confronted an array of novel and intricate obstacles during the transition to utilizing CF. The first obstacle initially presented itself in the form of frigid temperatures, while the second challenge emerged in the context of the CF tanks' (CFTs) substantial proportions.

Fuel	Jet A-1 (Kerosene)	Liquefied Methane–CH_4	Liquefied Hydrogen H_2
Density ρ, t/m^3	0.82	0.49	0.07
Mass calorific value q_m, MJ/kg	42.8	50	120
Volumetric calorific value q_v, MJ/L	35.1	24.5	8.4
Boiling temperature T_b, °C	180	−162	−253

Table 2.
Features of fuels utilized in gas turbine engines [33].

In order to prevent boiling, it is necessary to maintain a CFT temperature below −253°C and to ensure effective thermal insulation. Given the potential for elevated internal pressure within CFT, tanks are advised to be spherical or cylindrical in shape to reduce overall mass. The volumetric heat of combustion indicates that kerosene is comparable to LH at a ratio of 35.1–8.4, or 4.18, and with LNG at a ratio of 35.1 and 24.5, or 1.43. It can be reasonably inferred that if the CF technology was to achieve the identical energy efficiency as kerosene and the same flight performance for UAVs, the resulting increase in CFT volume would be comparable, and consequently, an enlargement of the aircraft would ensue. Therefore, an elevation in aerodynamic drag will consequently result in an increased rate of fuel consumption, thereby necessitating augmented engine thrust. As a result, the mass of the aircraft will also grow, consequently increasing the airframe mass. To maintain the Tu-154 aircraft's aerodynamic shape, CFTs (canopies) were installed in the rear fuselage compartment. This consequently led to a decrease of approximately 50% in the aircraft's permitted passenger capacity.

Since 1988, this aircraft has completed five flights. During this period, the developers encountered challenges in utilizing LH as a fuel source. It became evident that a more seamless transition between high-boiling and low-boiling fuels was necessary to gain experience. As a preliminary step, the Tu-155 underwent modifications to enable the use of the more cost-effective and concentrated CF [11]: LNG fuel, which encompasses a range of chemical compounds, including CH_4 (methane) and C_5H_{12} (pentane). A significant rationale for this modification was the necessity for LNG to be stored at temperatures below −162°C during transportation. Nevertheless, LNG is markedly less costly than not only LH but also approximately twofold less expensive than kerosene. Furthermore, LNG is less flammable than hydrogen. Consequently, by the time this fuel was first utilized, sufficient experience had been accumulated in maintaining it safely. Since 1989, LNG has completed 90 successful test flights, which demonstrated that fuel consumption is reduced by approximately 15% in comparison with kerosene. This value corresponds to the ratio of the specific heat capacities of LNG and kerosene, indicating that a cryogenic aircraft could become a more economically sustainable solution.

A fuel-economic analysis was conducted for the Tu-206 aircraft project [11]. The baseline model utilized in this analysis is the existing Tu-204 aircraft, which has been retrofitted with three Kuznetsov NK-89 engines, collectively capable of delivering a total thrust of 320 kN. The aircraft is equipped with two distinct fuel systems: a conventional kerosene system and a novel LNG system. The findings of this study indicate that despite an increase in the mass of the new aircraft by 13%, there is an annual economic benefit due to the reduction in LNG costs, amounting to 1.8% of the cost of the entire Tu-204 ($46,000,000). As demonstrated by the research [38], the efficiency of the hydrogen fuel in comparison with that of traditional fuel is not greater than 35%. The conversion to LNG can provide airlines with significant cost savings when operating passenger aircraft for 20–30 years. This reduction in costs can be passed on to passengers in the form of lower ticket prices, while also reducing the environmental impact.

The most challenging aspect of utilizing CF for drones, which are relatively diminutive, is the positioning and configuration of fuel tanks. Given the particular shape of these tanks, it is evident that their volume increases considerably in comparison with traditional hydrocarbon fuels. This gives rise to whether there is a case for installing and optimally placing CF fuel tanks on board the UAV.

The absence of comprehensive data introduces complexity into the design process, necessitating a conceptual approach that yields reliable and straightforward outcomes. In aviation, the most critical factor influencing an aircraft's performance is its take-off mass. A review of the literature on the potential use of CF for UAVs at the conceptual level reveals that the majority of listed researches do not consider the relationship between the use of CF and the take-off mass. This lack of consideration for the influence of the take-off mass on the design process is a significant limitation, as it fails to account for a primary factor in determining the viability of further design.

The objective of the proposed numerical research is twofold: firstly, to examine the sensitivity of the take-off mass to design changes for transitions from traditional fuel to CF and secondly, to investigate the conceptual possibility of using CF in UAVs. The present study employs the MQ-9 Reaper drone [17], a currently widely deployed aircraft with a take-off mass of up to five tons, developed by General Atomics Aeronautical Systems (USA), as a specific illustrative example. The aircraft is outfitted with a turboprop engine, enabling it to attain speeds in excess of 400 kilometers per hour and elevations above 13.2 kilometers during flight. The aircraft is capable of remaining aloft for up to 25 hours. In this study, the authors utilize numerical analysis to assess the feasibility of incorporating LH and LNG into metal and composite fuel tank designs. In order to evaluate the rationale for utilizing CF, the minimum take-off mass criterion is employed as a basis for comparison with the MQ-9 Reaper UAV, which is used as a baseline UAV.

4.2 Justification for transition to new design solutions

For any aircraft, including UAVs, the take-off mass and, in particular, the maximum mass is among the most critical design parameters. In aircraft design, knowledge of the maximum mass is necessary for a number of reasons. It is a key consideration when assessing the design loads, which largely determine the structure's mass. It also informs the assessment of the thrust-to-weight ratio of an aircraft, the selection of parameters for a power plant, and the determination of fuel costs. At the conceptual design stage, significant technical decisions can reach about 80%. Concurrently, developers must consider and sort through a multitude of options within a relatively limited timeframe. Therefore, it is advantageous to have an approach that is both straightforward in terms of labor intensity and sufficiently proven to enable the attainment of an effective design direction in a timely manner. In order to gain insights into the impact of changes in technical requirements on the development process, we adopted a methodological approach based on existing, albeit basic, projects.

In studies [39–46] have explored the potential of aircraft take-off mass growth factors due to initial changes as a solution to the aforementioned problems. As demonstrated in [47, 48], this approach can be regarded as a form of sensitivity analysis. A sensitivity coefficient is defined, in general terms, as the ratio of the change in the final value of a given parameter to its initial value. The assessment of the sensitivity coefficients of the base drone to new structural changes serves as the foundation for the proposed study.

In order to determine the take-off mass of the aircraft, it is necessary to consider the sum of the main functional components:

$$m_{\text{TO}} = \sum_{i=1}^{4} m_i = m_{\text{str}} + m_{\text{p.p}} + m_{\text{fuel.s}} + m_{\text{targ}}, \tag{1}$$

where m_{str} is the mass of the structure; this mass is calculated by combining all functional components into a single entity and ensuring that the payload is securely positioned within the structure; $m_{p.p}$ is used to denote the mass of the power plant that is defined as the subsystem responsible for thrust generation; this encompasses engines, nacelles, a frame with attachment points, and other associated components (it is acknowledged that this subsystem depends on the mass of engines in the following ways [47]: $m_{p.p} = k_{p.p} \times m_{eng}$, where m_{eng} is the mass of the engines, usually the factor $k_{p.p} \approx 1.1$); $m_{fuel.s}$ is the mass of the subsystem responsible for supplying fuel to the power plant over the course of a given flight; this subsystem incorporates fuel and the devices necessary for its placement and distribution, collectively representing a mass of between 10 and 15% of the total fuel mass; m_{target} is the mass of all components related with the intended purpose of the aircraft; for the purposes of this analysis, the payload, the crew, payload equipment, and the equipment necessary for reliable and safe flight will be considered a subsystem.

As a general rule, the fourth term on the right-hand side of eq. (1), m_{targ}, is primarily determined by the specific design task assigned to the aircraft in question, whereas the first three terms are directly dependent on m_{TO}. To address this issue, specific values are utilized to calculate the take-off mass for the preliminary approximation: $\overline{m}_i = m_i / m_{TO}$, thereby

$$m_{TO} = \frac{m_{targ}}{1 - \left(m_{str} + m_{p.p} + m_{fuel.s}\right)}. \tag{2}$$

The new solutions application will be implemented through the primary parameters of the aircraft, with the resulting alterations reflected in the m_{TO} of the original project. The transfer from an extant basic project to a novel one will require partial modifications to the mass of the functional elements of the system, which will then give rise to a comprehensive alteration in the configuration of the aircraft as a system constituted by the linkage of such elements. Modifications to the primary project's technical indicators, including $\overline{m}_{str}$, $\overline{m}_{p.p}$, and $\overline{m}_{fuel.s}$, will result in a shift in the take-off mass (m_{TO}), which can be regarded as a continuous function of *n* variables, including mass, aerodynamics, economics, and so forth. These parameters serve to determine the total mass differential, with finite but small changes to the parameters q_j taken into account. This transition can be represented as:

$$dm_{TO} = \sum_{j=1}^{n} \frac{\partial m_{TO}}{\partial q_j} dq_j \rightarrow \Delta m_{TO} = \sum_{j=1}^{n} \frac{\partial m_{TO}}{\partial q_j} \Delta q_j. \tag{3}$$

Consequently, any alterations to the assessment of the take-off mass, particularly in light of the initial changes in the related functional mass Δm_{i0} (i = 1—4, with the index "0" signifying the initial change in mass), then

$$\Delta m_{TO} = \frac{\partial m_{TO}}{\partial m_j} \Delta m_{i0} = \mu_{mi} \Delta m_{i0}, \tag{4}$$

where μ_{mi} is the aircraft's mass sensitivity coefficient (MSC), which can be defined as the ratio of the change in mass to the change in acceleration. Once the transition from infinitesimal quantities to finite increments has been made, it is possible to express the following:

$$\mu_{mi} = \frac{\partial m_{TO}}{\partial m_j} \approx \frac{\Delta m_{TO}}{\Delta m_{i0}}. \tag{5}$$

Without delving into the specifics of the MSC acquisition process, we direct the reader to [45, 47] for further information. In this study, the final value of the coefficient in question is presented with respect to a scenario in which there is an initial mass variation in the i-th functional section of the aircraft while other aircraft performance indexes continue to remain unaltered.

$$\mu_{mi} = \frac{1}{\left[\overline{m}_{\mathrm{targ}} - \Delta\overline{m}_{i0} + \left(\overline{m}_{\mathrm{p.p}} + \overline{m}_{\mathrm{fuel.s}}\right)\frac{C_{\mathrm{Dfus}}}{C_{\mathrm{D}}}\right]}, \tag{6}$$

where $\Delta\overline{m}_{i0} = \Delta m_{i0}/m_i$ (in order to streamline the problem, the specific mass, which introduces relatively minor corrections, will not be taken into further consideration); the drag coefficients C_{Dfus} and C_{D} refer to the fuselage and the aircraft as a whole, respectively. In the previous iteration of this analysis, the coefficients in eq. (6) were employed in a scenario wherein the fuselage dimensions were maintained constant, whereas the payload mass was adjusted accordingly [45]. In the event of a change in the dimensions of the fuselage, this term is no longer applicable.

The final equations for estimation of the masses for the new project can be expressed as follows:

$$\Delta m_{\mathrm{TO}} = \sum_{i=1}^{4} \mu_{m\,i} \Delta m_{i0}, \tag{7}$$

$$\Delta m_i = \Delta m_{i0} + \overline{m}_i \Delta m_{\mathrm{TO}}, \text{for } i = 1\text{—}3, \tag{8}$$

$$\Delta m_{\mathrm{targ}} = \Delta m_{\mathrm{targ0}}, \text{for } i = 4. \tag{9}$$

The use of linearization in the calculation of the MSC assumes an initial change in the functional masses due to design changes of no more than 10—15% (**Figure** 7) [47]. In this context, the parameter $\Delta\overline{m}_{i0}$ in Eq. (6) can be neglected and then for all functional masses $\mu_{mi} = \mu_m$.

A graphical illustration of the MSC with characteristic points is presented in **Figure** 7, here: 1—take-off mass of the base aircraft; 2—initial change of the target load mass while maintaining the take-off mass by changing the nominal fuel mass, and vice versa; 3—initial change of the target load mass without affecting the other functional masses with the corresponding loss of performance; 4—exact (nonlinear) value of the take-off mass of the new aircraft while changing the mass of the i-th functional mass and maintaining the flight characteristics of the aircraft; 5—approximate (linear) value of the take-off mass of the new aircraft obtained on the basis of the sensitivity coefficient.

4.3 Numerical research

The potential for transferring the UAV to the CF will be examined at the conceptual level. A similar study was carried out for the Blended Wing Body aircraft [49, 50]. The MQ-9 Reaper, a medium-altitude aircraft with a long flight duration, is selected as the base case for numerical analysis purposes. The MQ-9 Reaper represents the primary drone utilized by the United States Air Force for conducting strikes. Three projections of the MQ-9 Reaper, together with the average dimensions, are shown in **Figure** 8 [17].

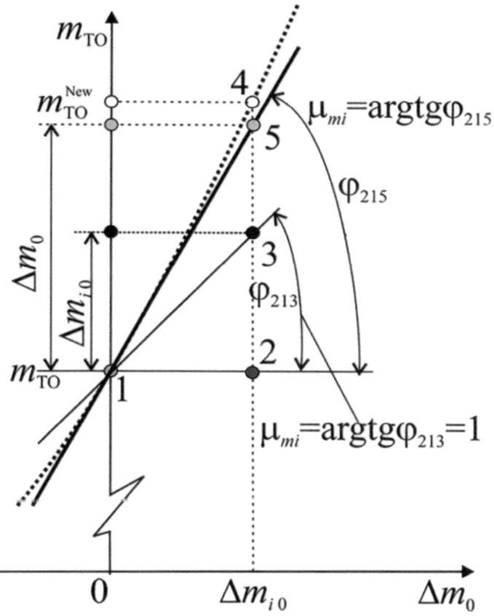

Figure 7.
Graphical explanation of the MSC [47].

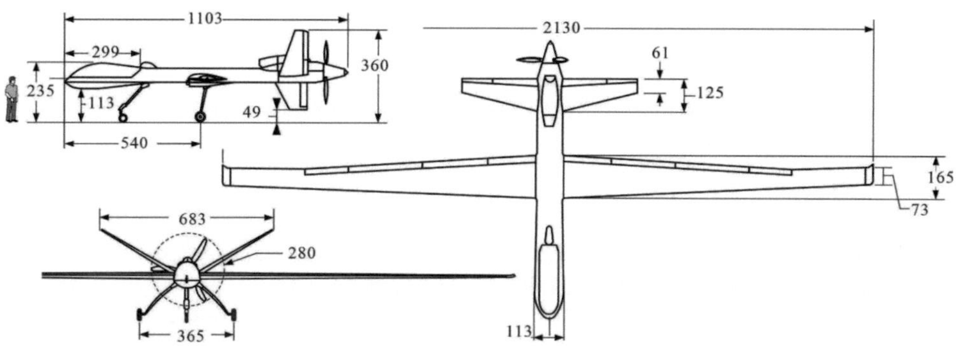

Figure 8.
MQ-9 Reaper three projections, all measurements are in centimeters [17].

The initial step is to determine the functional mass of the base aircraft, which is essential for calculating the MSC. The Honeywell TPE331-10 T turboprop engine is used in this UAV. It has an output power of 712 kW, a mass of 175 kg, a specific fuel consumption linked to its power of c_N = 325 g/(kWh). The mass of the power plant consists of a gearbox, propeller, and automation, in addition to the aforementioned engine. The component of the vehicle that bears the greatest weight, subsequent to the engine, is the gearbox. The statistical correlation between the gearbox's mass and the transmitted power is shown in **Figure 9** [51]. In light of the aforementioned evidence, it seems reasonable to posit the following dependence [9]: $m_{p.p} = 1.7 \times m_{eng} = 300$ kg, so, $\overline{m}_{eng} = 0.062$.

It is essential to ascertain the functional masses of the base aircraft in order to facilitate the MSC determination. In order to estimate the mass of the airframe, the data provided in Ref. [52] will be utilized. The dependency diagram demonstrating the relationship between the m_{str} and the specific wing load p for an airframe incorporating a considerable use of composite materials is illustrated in **Figure 10**. The wing area

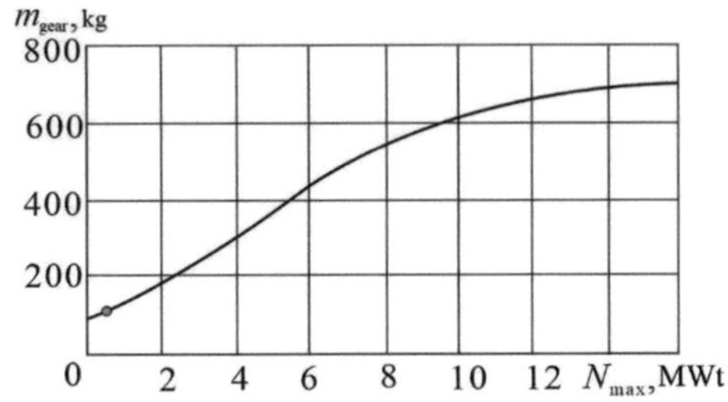

Figure 9.
Correlation between the gearbox's mass and the transmitted power [51].

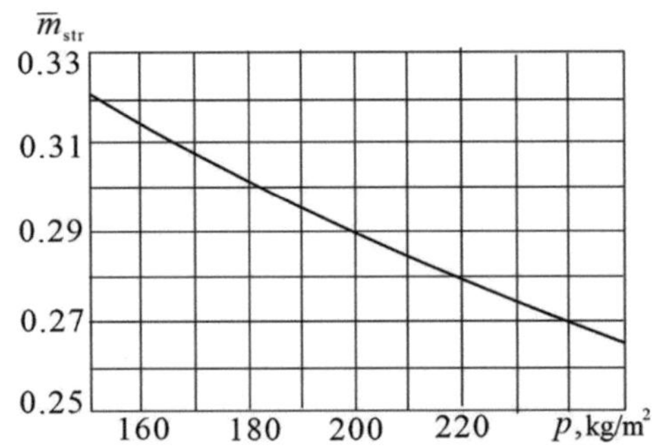

Figure 10.
Diagram of the relationship between the airframe's mass and the specific wing load [53].

is S = 25.4 m^2 and p = 1874 N/m^2 for a base aircraft. Thus, $\overline{m}_{str}$ = 0.29 and m_{str} = 1389 kg.

For the maximum mass of fuel (m_{fuel} = 1800 kg), the mass of the whole fuel system can be assumed as [47] $m_{fuel.s} = k_{fuel.s} \times m_{fuel}$. For $k_{fuel.s}$ = 1.055, the fuel system mass is $m_{fuel.s}$ = 1900 kg, and it is meant that $\overline{m}_{fuel.s}$ = 1900/4750 = 0.4. For the option with the maximum fuel mass, the payload mass is determined by the following equation [47]:

$m_{p.l} = m_{TO} - m_{emp} - m_{fuel}$ = 4760–2223 – 1800 = 737 kg.

By determining the empty mass of the aircraft, which includes the mass of the structural components, the power plant, the dry fuel system, and non-removable equipment, it is possible to calculate the mass of the installed equipment m_{eq}. As previously noted, this mass is assessed to be independent of the take-off [47]:

$m_{eq} = m_{emp} - (m_{str} + m_{p.p} + m_{fuel.s.emp})$ = 2223 – (1389 + 300 + 100) = 434 kg.

Consequently, the target load's mass can be calculated as follows [47]:

$m_{targ} = m_{eq} + m_{p.l}$ = 434 + 737 = 1171 kg.

The preliminary analysis yielded the following specific functional masses for the base aircraft: $\overline{m}_{str}$ = 0.3; $\overline{m}_{p.p}$ = 0.06; $\overline{m}_{fuel.s}$ = 0.4; $\overline{m}_{targ}$ = 0.24.

The initial change in mass is MSC, contingent upon the aerodynamic parameters of the fuselage and the established ratio remaining unaltered. According to Eq. (6) and $C_{D\ fus}/C_D$ = 0.3, it has view:

$$\mu_m = \frac{1}{0.24 + (0.4 + 0.06) \times 0.3} = 2.65$$

A change in the size of the fuselage may be occasioned by rearrangement for high MSC. Thus, μ_m = 1/0.24 = 4.2.

It is assumed that the transition to CF will not affect the mass of the turboprop engine; however, the traction properties will undergo a proportional alteration in relation to the mass heat of combustion. The mass of CF can be determined as follows:

- For LNG: $m_{\text{fuel LNG}} = q_{m\ \text{LNG}}/q_{m\ \text{mfuel}} = 50/42.8 \times 1.8 = 1.54$ t;
- For hydrogen: $m_{\text{fuel LH}} = q_{m\ \text{LH}}/q_{m\ \text{mfuel}} = 120/42.8 \times 1.8 = 0.64$ t;
- As a result, the initial change in the quantity of fuel will be as follows:
- For LNG: $\Delta m_{\text{fuel LNG 0}} = -0.26$ t;
- For hydrogen: $\Delta m_{\text{fuel LH 0}} = -1.16$ t.

Following the calculation of the necessary volumes of CFT and the subsequent increase by 15% (in consideration of the requisite free volumes for refueling and fuel fittings), the results are as follows: $W_{\text{LNG}} = 3.6$ m^3, and $W_{\text{LH}} = 10.5$ m^3.

The potential for installing CFT on UAVs will be evaluated. In the case of cylindrical CFTs with a radius of r, spherical bases with a radius of R are utilized. In order to guarantee the equivalent strength of the MSC and to reduce its mass, the relation for $R = 2 \times r$ is correct. Thus, the basis's removal will be $h = (2–(3)^{0.5}) \times r = 0.27 \times r$. The volume that forms the base has a spherical section with a radius of R, a height of h, and a radius of r:

$$W_{\text{bott}} = \pi \times h^2(R–h/3) = \pi \times 0.14 \times r^2 \times (2 \times r–0.09 \times r) = \pi \times 0.27\, r^3 \tag{10}$$

Consequently, the total volume of the tank can be expressed as:

$$W_{\text{CFT}} = W_{\text{CFTcyl}} + 2 \times W_{\text{CFTbott}} = \pi \times r^2 \times (l_{\text{CFT}}–0.54 \times r) \tag{11}$$

As illustrated in **Figure 11**, there is a correlation between the dimensions of the CFT, including length, radius, and the required volumes.

In accordance with the specifications set forth by the CFT layout conditions on the fuselage, the following values were selected: for LNG $r_{\text{CFT}} = 0.55$ m, $l_{\text{CFT}} = 4$ m; for LH $r_{\text{CFT}} = 0.75$ m, $l_{\text{CFT F}} = 6.4$ m.

A connection will be established between the CFT and the base UAV. In consideration of the aircraft's intended use for civilian purposes, it can be reasonably assumed that modifications to the forward fuselage may be feasible. When installing CFT with LNG (**Figures 12, 13**), the midship section—which in this instance has the same dimensions as the nose of the MQ-9 Reaper—is extended to the point where the wing attaches to the fuselage (**Figure 13**).

The configuration of the tank in question will guarantee the maintenance of the power scheme for the attachment of the wing to the fuselage.

In the case of CFT with hydrogen, it will be necessary to increase the diameter of the fuselage (see **Figures 13, 14**). Furthermore, the implementation of CFT in a single-tank scheme will result in the disruption of the wing's center section. In order to remedy this, power frames must be incorporated for the wing sections' attachment, and supplementary main landing gear's compartments must be added. The implementation of this scheme will result in an increase in aerodynamic drag and an increase in the structural mass. Moreover, the fuselage, which has a substantial

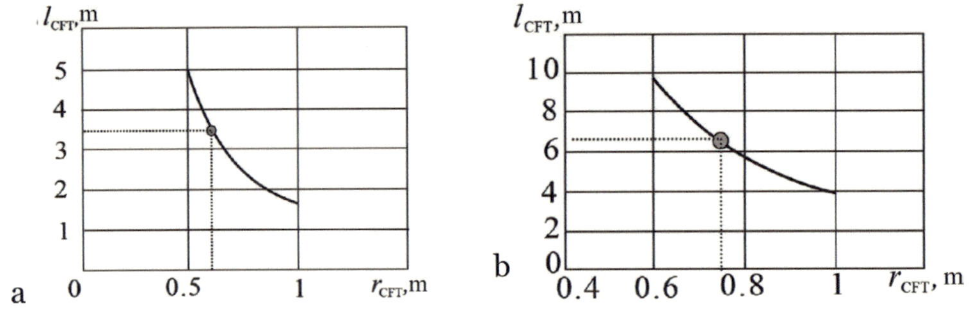

Figure 11.
Correlation between the length of CFT and the radius: (a) – for LNG (volume 3.6 m^3) (b) – for LH (volume 10.5 m^3) [17].

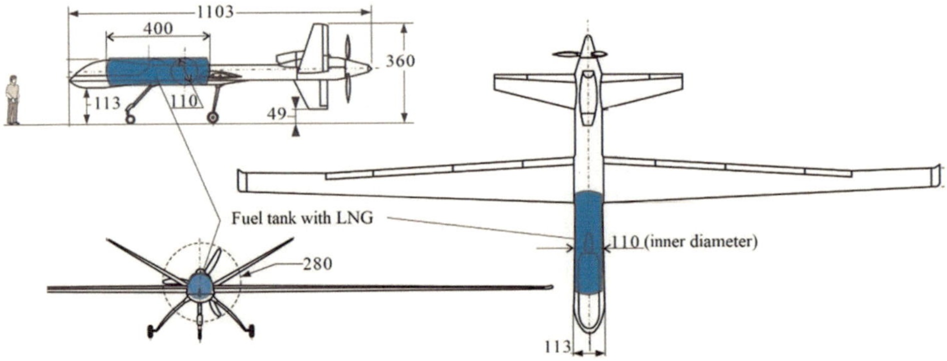

Figure 12.
Initial variant of CFT installation with LNG for a basic UAV (all measurements are in centimeters).

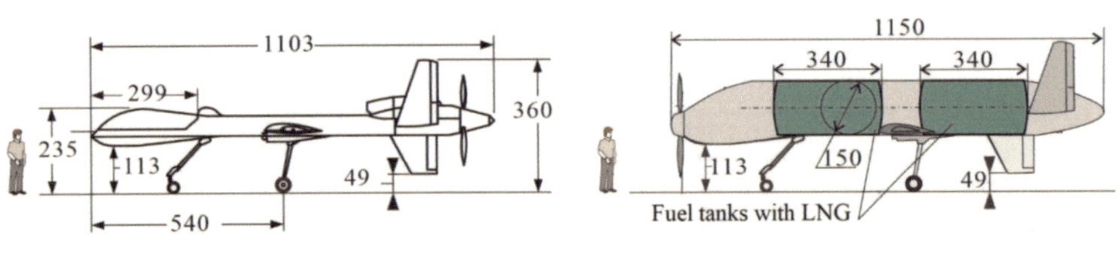

Figure 13.
Basic UAV vs. the twin-tank variant on LNG with the reconfigured power plant (all measurements are in centimeters).

midsection, will provide a considerable degree of shading for the propeller. Consequently, it will be necessary to relocate the power plant from its current position at the rear of the fuselage to the nose, thereby utilizing a propeller group that pulls rather than pushes. In this scenario, it would be more logical to construct two CFTs. The length of each 1.5-meter inner-diameter tank is 3.4 meters.

It is of the utmost importance to integrate the tank's thermal protection system with the fuel system, taking into account the mass of the tank itself. It is presumed that the preexisting tanks for kerosene, equipped with a system for supplying it to the engine, compensate for the mass of the cryogenic component supply system.

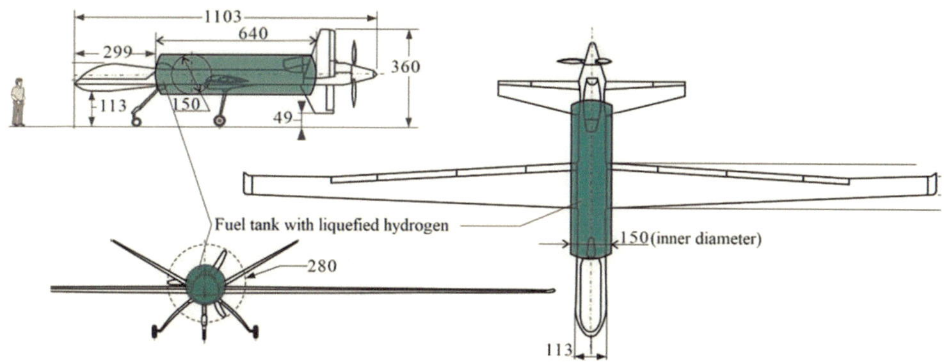

Figure 14.
Initial scheme of CFT with LH on the base UAV with a single-tank (all measurements are in centimeters).

In order to obtain an estimate of the mass of CFT without delving into its structural specifics, it is possible to use the results of [52]. In this study, the mass of a thin-walled cylindrical airframe of a circle cross-section was determined for various internal pressure p, using the fourth theory of strength. The following relations were considered:

- For an isotropic structural material (metal):

$$m_{\text{CFT }0} = 3^{0.5}\frac{pW}{\overline{\sigma}_{\text{met}}} \tag{12}$$

- For an anisotropic structural material (composite):

$$m_{\text{CFT }0} = 3^{0.5}\frac{pW}{\overline{\sigma}_{\text{CM}}} \tag{13}$$

where $\overline{\sigma}$ is the specific strength of the structural material, represents the ratio of given stresses to the density [6], $\overline{\sigma} = \sigma_B/\eta/\eta_p/\rho$ (η is the entire factor of safety, and η_p is the factor of safety related with operation under overpressure). For composite materials, an extra factor of safety, denoted as η_{CM}, should be implemented.

In order to calculate the initial mass change for fuel and the power plant related to the growth of aerodynamic drag for a cruise flight mode, it is possible to take the following relationship into account

$$\Delta D_0 = \Delta T = \frac{T}{\left(m_{\text{eng.s}} + m_{\text{fuel.s}}\right) \times \left(\Delta m_{\text{eng.s0}} + \Delta m_{\text{fuel.s0}}\right)} \tag{14}$$

Thus, it follows:

$$\Delta D_0 = \frac{g}{K \times \left(\overline{m}_{\text{eng.s}} + \overline{m}_{\text{fuel.s}}\right) \times \left(\Delta m_{\text{eng.s0}} + \Delta m_{\text{fuel.s0}}\right)} \tag{15}$$

In accordance with the established definition, the CFT in terms of aerodynamic drag is:

$$\mu_D = \frac{\Delta m_{\text{TO}}}{\Delta D_0} = \mu_m \times K \times \frac{\overline{m}_{\text{eng.s}} + \overline{m}_{\text{fuel.s}}}{g} \tag{16}$$

To determine the impact of an alteration in the diameter of the mid-section of the fuselage ($d_{\text{fus}} \Rightarrow d_{\text{fus New}}$) on the force of aerodynamic drag, we postulate that the drag exerted by the fuselage is directly proportional to its mid-section area, S_{fus}, and thus:

$$\frac{D_{\text{fus New}}}{D_{\text{fus}}} = \frac{S_{\text{fus New}}}{S_{\text{fus}}} = \left(\frac{d_{\text{fus New}}}{d_{\text{fus}}}\right)^2 \tag{17}$$

It is estimated that the average value of the aerodynamic drag of the base drone is approximately half the fuel consumption.

$$D_{\text{fus}} = 0.3D = \frac{0.3 \times (m_{\text{TO}} - 0.5 \times m_{\text{fuel}}) \times g}{K} \tag{18}$$

Subsequently, the alteration in the drag force of the fuselage for novel iterations of UAVs may be quantified by employing the following equation, assuming that the drag coefficient remains invariant:

$$\Delta D_{\text{fus}} = D_{\text{fus}} \times \left(\left(\frac{d_{\text{fus New}}}{d_{\text{fus}}}\right)^2 - 1\right) \tag{19}$$

where $d_{\text{fus New}}$ refers to the fuselage's diameter of its midsection of the new UAV; this value is determined by employing an integrated methodology that incorporates the overall dimensions of the CFT and accounts for the heat-shielding coating.

The current study assesses the perceived values of two distinct variants of fuel tanks: aluminum alloys and composite materials. The calculations were performed using specific parameters: η = 1.5; η_p = 3.5; η_{CM} = 1.5; $\sigma_{\text{B alum all}}$ = 600 MPa; $\sigma_{\text{B CM}}$ = 1500 MPa; $\rho_{\text{alum all}}$ = 2.7 t/m^3; ρ_{CM} = 1.5 t/m^3; p = 1 MPa. In order to achieve thermal protection for the CFT, a polyurethane foam coating with a purge system was considered, with the total density of the system being a key factor in the evaluation process $\rho_{\text{thermal insulation}}$ = 100 kg/m^3; the thickness of the heat-protective coating (HPC): for CFT with LNG δ_{HPC} = 0.1 m, for CFT with LH δ_{HPC} = 0.2 m. These parameters were accepted for the base drone: diameter of the fuselage mid-section d_{fus} = 1.13 m; lift-to-drag ratio is K = 25.

In accordance with the accepted values of the parameters and the eq. (18), the drag of the fuselage of the base UAV in cruise flight is D_{fus} = 454 N.

Table 3 illustrates the findings of a conceptual investigation into the potential for modifying the basic UAV (MQ-9 Reaper) into the scheme with CF.

In order to ascertain the revised functional mass values, Eq. (8) is utilized. The results of the calculations are presented in tabular form in **Table 3**.

A review of the data of Δm_{TO} in the final row of **Table 3** reveals that for a UAV with a take-off mass of approximately 5 t, the use of LH is not a rational choice. This phenomenon can be attributed to an expansion in the transverse dimensions of the fuselage, which is designed to accommodate CFT. This modification markedly increases the aerodynamic drag and the corresponding fuel consumption. The use of LNG results in an approximate 6–8% increase in take-off mass. Consequently, a comprehensive evaluation of the rationale for employing such fuel is essential,

Type of fuel for the UAV	LNG		LH	
CFT material	**Aluminum alloys**	**Composite materials**	**Aluminum alloys**	**Composite materials**
MSC μ_m	2.65	2.65	4.2	4.2
$\Delta m_{\text{fuel 0}}$, t	−0.26	−0.26	−1.16	−1.16
$\Delta m_{\text{CFT 0}}$, t	0.085	0.047	0.46	0.14
$\Delta m_{\text{HPC 0}}$, t	0.16	0.16	0.67	0.67
$\Delta m_{\text{TO 1}} = \mu_m(\Delta m_{\text{fuel 0}} + \Delta m_{\text{CFT 0}} + \Delta m_{\text{HPC 0}})$, t	−0.039	−0.140	−0.126	−1.47
d_{fus}, m	1.3	1.3	1.9	1.9
$\Delta D_{\text{fus 0}}$, N	146	146	829	829
$\Delta m_{\text{aer 0}} = \Delta m_{\text{fuel 0}} + \Delta m_{\text{p.p 0}}$, t	0.172	0.172	0.973	0.973
$\Delta m_{\text{TO aer}} = \Delta\mu_m \Delta m_{\text{aer 0}}$, t	0.456	0.456	4.087	4.087
$\Delta m_{\text{TO}} = \Delta m_{\text{TO 1}} + \Delta m_{\text{TO aer}}$, t	0.417	0.316	3.961	2.618
$(\Delta m_{\text{TO}}/m_{\text{TO}})\cdot 100\%$	8.76	6.64	83.2	54.8

Table 3.
Assessment of the mass changes of LNG and LH UAV with two types of fuel tanks.

encompassing a meticulous economic analysis of the design scenario and a more precise assessment of all design parameters, including data on fuel, operational costs, the specifics of UAV operation, and other pertinent factors.

5. Conclusions

The methodologies employed for the implementation of the ICAO requirements were elucidated and subjected to critical analysis. A brief overview of the history of UAVs and hydrogen technologies was provided. Additionally, the potential for integrating these technologies to enhance their respective benefits was explored. This analysis was conducted using a conceptual methodology take-off mass sensitivity analysis due to design changes in the base UAV (in this case, the transition from hydrocarbon fuel to CF).

Transition from hydrocarbon fuel to environmentally friendly fuel based on LH is quite a big leap that requires a lot of new design and technological solutions. As the experience of creating a new aircraft based on Tu-154 variants of Tu-155 and Tu-156 has shown, it is recommended to first pass through an intermediate stage—the use of LNG—in order to switch to LH, which will allow to accumulate the necessary experience for the use of a truly green fuel.

In light of the prospective utilization of liquefied gases (methane and hydrogen), it is imperative to undertake a preliminary assessment of the impact of modifications to the design of UAVs. The subject of the study was the MQ-9 Reaper drone. The results of the calculations demonstrated that for drones of this category when utilizing LNG, the take-off mass exhibited an increase of approximately 6—8%. It is thus advised that a comprehensive economic and environmental analysis be conducted, assessing all pertinent design parameters with precision, including data pertaining to fuel, operational costs, the distinctive operational characteristics of the UAV in question, and so forth. A brief examination of the mass analysis does not provide sufficient grounds to conclude that the use of LH for this drone is unwarranted.

The substantial increase in the volume of fuel tanks with CF, due to their low density and high internal pressure, necessitates the implementation of a UAV scheme BWB, which allows for the accommodation of tanks of spherical or cylindrical shape within its large internal volume. This is in accordance with the findings presented in [47], which demonstrate the suitability of this approach for mainline passenger aircraft.

Author details

Longbiao Li[1], Anatolii Kretov[2] and Dmytro Tiniakov[1*]

1 Nanjing University of Aeronautics and Astronautics, Nanjing, China

2 Moscow Aviation Institute, National Research University, Moscow, Russia

*Address all correspondence to: tiniakov_d@nuaa.edu.cn

References

[1] Crabtree T. Air Cargo Traffic, Capacity and Revenues Continued to Recover in Q4. Trade and Transport Group: Sydney; 2024. Available from: https://www.aircargonews.net/airlines/air-cargo-traffic-capacity-and-revenues-continued-to-recover-in-q4/ [Accessed: June 1, 2024]

[2] Grossman A, Haywood J, Rind D, Subbaraya B. Aviation and the Global Atmosphere. Geneva: IPCC; 2000. Available from: https://archive.ipcc.ch/ipccreports/sres/aviation/index.php?idp=63 [Accessed: June 1, 2024]

[3] Gielen D, Oksanen S. Advanced Aviation Biofuels: Ready for Take-off? Amsterdam: Energypost.eu; 2019. Available from: https://energypost.eu/advanced-aviation-biofuels-ready-for-take-off/ [Accessed: June 1, 2024]

[4] Zhang F, Maddy J. Investigation of the Challenges and Issues of Hydrogen and Hydrogen Fuel Cell Applications in Aviation. Missuri: TechRxiv; 2021. Available from: https://www.techrxiv.org/doi/full/10.36227/techrxiv.14958057.v1 [Accessed: September 16, 2024]

[5] Tinyakov DV. Analysis of geometric parameters features of the wings of modern transport category airplanes from the top-view. Open Information and Computer Integrated Technologies. 2013;**62**:14-20

[6] Kretov AS, Tiniakov DV, Shataev PA. Conceptual assessment of the fuel efficiency of passenger aircraft with the transition to composite wings. Journal of Civil Aviation High Technologies. 2023; **26**(2):72-90. DOI: 10.26467/2079-0619-2023-26-2-72-90

[7] Li L, Tiniakov D. Propulsion Systems —Recent Advances, New Perspectives and Applications. Chapter: Review of the Aviation Power Units Development. London: IntechOpen; 2023. DOI: 10.5772/intechopen.112741

[8] Tiniakov D, Li L, Su Y. Airworthiness Design of Civil Aircraft Systems and Structures. Beijing: Beijing Press; 2018. p. 363

[9] Bicer Y, Dincer I. Life cycle evaluation of hydrogen and other potential fuels for aircrafts. International Journal of Hydrogen Energy. 2017; **42**(16):10722-10738

[10] Tupolev AA. Clear sky variant. Technics of Youth. 1989;**1**:18-21

[11] Kretov AS, Glukhov VV. Alternative fuel in transport aviation and estimation of its application efficiency. Journal of Russian Aeronautics. 2021;**64**(3): 365-375. DOI: 10.3103/S1068799821030016

[12] Boukoberine MN, Zhou Z, Benbouzid M. A critical review on unmanned aerial vehicles power supply and energy management: Solutions, strategies, and prospects. Applied Energy. 2019;**255**(113823):1-22

[13] Hodgkinson D, Johnston R. Aviation Law and Drones. New York: Routledge; 2018. p. 176. ISBN: 978-1-351-33232-3

[14] Mason K. The British Bomber Since 1914. London: Putnam; 1994. ISBN 0-85177-861-5

[15] John O. The V1, the Machine and its Men. North Charleston: CreateSpace Independet Publishing Platform; 2018. ISBN 978-1-987754-75-9

[16] Thomas EP. Air Force UAVs: The Secret History. Defense Technical

Information Center (DTIC(R)). Arlington: Mitchell Institute for Airpower Studies; 2015

[17] Zaloga SJ. Unmanned Aerial Vehicles: Robotic Air Warfare 1917–2007. London: Bloomsbury Publishing; 2011. ISBN 978-1-84603-786-3

[18] Brar R. Design of a Blended Wing Body Aircraft. SanJose: SanJose State University; 2014. p. 99

[19] Maddalon J, Hayhurst K, Koppen D, et al. Perspectives on Unmanned Aircraft Classification for Civil Airworthiness Standards. Virginia: NASA; 2013

[20] Fahlstrom P, Gleason T. Introduction to UAV Systems. West Sussex: WILEY; 2012. p. 287. ISBN: 978-1-119-97866-4

[21] Sebbane Y. Intelligent Autonomy of UAVs. New York: CRC Press; 2018. p. 421. ISBN: 978-1-138-56849-5

[22] Tiniakov D, Li L. Airworthiness Engineering Basis. Beijing: Beijing Press; 2020. 217 p. ISBN 9787512432093

[23] Kirdyushkin Y. Potential of hydrogen fuel application in civil aviation. Moscow: MGTU GA. Journal of Civil Aviation High Technologies. 2013; **194**:110-113

[24] Bruce S, Temminghoff M, Hayward J, Palfreyman D, Munnings C, Burke N, et al. Opportunities for Hydrogen in Aviation. Canbera: CSIRO. Commonwealth Scientific and Industrial Research Organization; 2020

[25] Tsiolkovsky K. Proceedings on Rocket Technology. Moscow: URSS/LIBROCOM; 2009. p. 368

[26] Braun H. The phoenix project: Shifting to a solar hydrogen economy by 2020. CI&CEQ. Journal Chemical Industry and Chemical Engineering Quarterly. 2008;**14**(2):107-118. DOI: 10.2298/CICEQ0802107B

[27] NASA. Atlas-Centaur AC-17 Performance for Applications Technology Satellite Ats-D Mission. Cleveland: NASA Lewis Research Center; 1972

[28] Bart H, Bert V. Energiya-Buran: The Soviet Space Shuttle. London: Springer; 2007. ISBN 978-0-387-69848-9

[29] Ozoroski A, Nickol L, Guynn D. High Altitude Long Endurance UAV Analysis Model Development and Application Study Comparing Solar Powered Airplane and Airship Station-Keeping Capabilities (No. L-20515). Hampton: NASA; 2015. Available from: https://ntrs.nasa.gov/citations/20150001258 [Accessed: June 1, 2024]

[30] Guynn M, Freeh J, Olson E. Evaluation of a Hydrogen Fuel Cell Powered Blended-Wing-Body Aircraft Concept for Reduced Noise and Emissions. Washington: NASA STI; 2004

[31] Swider-Lyons K, Stroman R, Page G, et al. The ion tiger fuel cell unmanned air vehicle. In: Las Vegas Conference: 44th Power Sources Conference. Washington: US Naval Research Laboratory; 2010

[32] Smart M. HyperSonix Launch Vehicles. Town Hall: AIAA/ASCEND; 2021

[33] Kretov A, Glukhov V, Tikhonov A. Conceptual assessment of the possibility of using cryogenic fuel on unmanned aerial vehicles. Journal of Drones. 2021;**6**: 217. DOI: 10.3390/drones6080217

[34] Stroman O, Schuette W, Swider-Lyons K, et al. Liquid hydrogen fuel

system design and demonstration in a small long endurance air vehicle. International Journal of Hydrogen Energy. 2014;**39**:11279-11290

[35] Sullivan M, Palko L, Tornabene T, et al. Engineering Analysis Studies for Preliminary Design of Lightweight Cryogenic Hydrogen Tanks in UAV Applications (No.E-15431). Hampton: NASA; 2006. Available from: https://ntrs.nasa.gov/citations/20060021606 [Accessed: June 1, 2024]

[36] Millis G, Tornabene T, Jurns M, et al. Hydrogen Fuel System Design Trades for High-Altitude Long-Endurance Remotely-Operated Aircraft (No. E-16800). Hampton: NASA; 2009. Available from: https://ntrs.nasa.gov/citations/20090013674 [Accessed: June 1, 2024]

[37] Verstraete D. Long range transport aircraft using hydrogen fuel. International Journal of Hydrogen Energy. 2013;**38**:14824-14831

[38] Kraev V, Tikhonov A, Kuzmina-Merlinoi I. Economic and ecological aspects of the use of new cryogenic aviation fuels. Journal of Applied Engineering Science. 2022;**20**(2):351-357. DOI: 10.5937/jaes0-31570

[39] Ballhaus WF. Clear Design Thinking Using the Aircraft Growth Factor. Vol. 540148. Warrendale: SAE; 1954. p. 15

[40] Drakin II. Influence of changes in the weight and aerodynamic characteristics of the structure on the flight weight of the aircraft. Moscow: Nauchnyi Vestnik MGTU GA. 1960;**1**: 52-62

[41] Politkovsky V I, Badyagin A A. On the coefficient of increase in the starting weight of aircraft. Moscow: Nauchnyi Vestnik MGTU GA, 1. 1966. pp. 161-164

[42] Ross H. Effect of the location of a fixed weight penalty on the aircraft growth factor. In: Proceedings of the 28th Annual Conference. San Francisco: University of San Francisco; 1969

[43] Badyagin AA. Aircraft Design. Moscow: Mashinostroenie; 1972. p. 516

[44] Roskam J. Airplane Design Part V: Component Weight Estimation. Ottawa: Roskam Aviation and Engineering Corp; 1985. p. 612

[45] Gogolin VP. Determination of the growth factor of takeoff weight changes in the implementation of changes in the weight of the structure. Kazan: Journal Aviation Technics. 1973;**160**:11-15

[46] Gogolin VP. On the problem of resolving contradictions between the weight and drag of aircraft parts. Kazan: Journal Aviation Technics. 1974;**1**:13-16

[47] Kretov A. Sensitivity factors of aircraft mass for the conceptual design. Moscow: Jornal Aircraft Engineering and Aerospace Technology. 2021;**93**: 1470-1477

[48] Kretov A, Tiniakov D. Evaluation of the mass and aerodynamic efficiency of a high aspect ratio wing for prospective passenger aircraft. Basel: MDPI Aerospace. 2022;**9**:1-17. DOI: 10.3390/aerospace9090497

[49] Kretov AS, Glukhov VV. Application of integrated layout for "cryogenic" transport category aircraft. Moscow: J. Russian Aeronautics. 2022;**65**:10-24

[50] Kretov A, Glukhov V, Tikhonov A. Conceptual assessment of the possibility of using cryogenic fuel on unmanned

aerial vehicles. Basel: MDPI Drones. 2022;**6**:1-15. DOI: 10.3390/drones6080217

[51] Grigoriev VA, Zagrebelny AO, Prokaev AS, Kuznetsov SP. On the issue of estimating the mass of the gearbox in the problem of optimizing the parameters of the working process of the TVB at the stage of initial design. Samara: Journal Vestnik of Samara University Aerospace and Mechanical Engineering. 2014;**4**:132-138

[52] Kretov AS, Shataev PA. Preliminary assessment of the mass of the aircraft fuselage as a result of the transition to composite materials. Moscow: Journal Russian Aeronautics. 2020;**63**:386-396

[53] Trokhov DA, Turkin IK. Formation of the appearance of a high-altitude unmanned aerial reconnaissance vehicle in search tasks. Moscow: Journal Civil Aviation High Technologies. 2015;**221**:106-114

Chapter 2

Methodology for Composite Blades Testing: From UAVs to Mega-Large Structures

Goran Vorotović, Jela Vorotović, Nebojša Petrović, Nikola Davidović, Đorđe Novković and Pavle Petrović

Abstract

The use of composite rotor blades in the domains of micro-aircraft, aircraft for the transport of goods and passengers, for military purposes, as well as in the domain of wind turbines represents one of the most significant aspects in the problems of aerodynamics and energy efficiency, especially from the aspect of safety. The real use of composite blades whose traction-dynamic, lift, and thrust dynamic characteristics in real conditions of exploitation are directly dependent on complex loads of a static and dynamic nature requires a comprehensive laboratory analysis according to appropriate methodological principles. This chapter presents the results of multi-decade research aimed at identifying and verifying a unique methodology for testing composite blades on examples of UAV blades, blades exposed to extreme conditions of exploitation, as well as blades used for electricity generation. It is interesting to note that the methodology, regardless of the dimensions of the blades and their purpose, is practically applicable to all cases of blade use in the earth's atmosphere, where obtained test results realistically describe the complex behavior of the blades in real exploitation conditions.

Keywords: propeller, traction-dynamic characteristics, oscillations, acquisition, dynamics

1. Introduction

Composite materials represent a combination of two or more materials with different physical and chemical properties on a macroscopic scale, with the aim of forming a new usable material. Composite, in fact, is a material consisting of an elastic matrix, which ensures the homogeneity and workability of the composite and filling, which ensures the rigidity and strength of the structure. The characteristics of the matrix are good shear resistance, low specific gravity, and good workability. The characteristics of the filling (reinforcing fibers) are high strength, high modulus of elasticity, and low specific weight. Based on these properties, we can come to the specifics of the composite, namely: strength much higher than the strength of the matrix, modulus of elasticity much higher than the modulus of the matrix, and specific weight approximately the same as the weight of the matrix.

It has been shown that the use of composite materials, for the supporting elements of the construction of aviation structures, provides a number of advantages compared to classic aviation materials. These advantages can be in terms of carrying capacity and construction:

- the transmission of forces through a larger number of fibers and the application of simultaneous polymerization of several structural elements in one tool enables obtaining a tough construction, which has a relatively high load capacity even after partial damage,
- the basic structure can be made from a small number of assemblies and subassemblies,
- in accordance with the previous one, the number of mechanical connecting elements can be reduced;

or in terms of servicing and exploitation:

- composites have high anti-corrosion properties,
- have low thermal conductivity,
- the finished surfaces are aerodynamically clean and do not require subsequent treatment,
- the process of overhauling the structure is quite simple and does not require special conditions [1, 2].

Composite materials, in addition to high strength and rigidity, are characterized by fatigue and vibration resistance. These are properties that directly affect the extension of the exploitation life of elements made of these materials. Composites are characterized by the retention of a high percentage of static strength, both unidirectional and transversely stacked laminates. In addition to all this, they are characterized by low density, which directly affects the reduction of the mass of the structure. How much this reduction will be depending on the complexity of the structure, its dimensions, and the intensity of the load.

The design of rotor blades is a balanced integration of aerodynamics, structural analysis and dynamics, choice of different types of composite materials, production technology, and overall economy.

The design of modern rotor includes choice of blade number, airfoil, chord, and twist distribution. Additional criteria are reliability, noise, and esthetic considerations. One of the important requirements when designing a propeller is the maximum reliability of the structure in all modes of exploitation. Such optimization requires the inclusion of all parameters through aerodynamic and structural modeling and the adoption of all major components of the system that includes the use of propellers.

Following is the methodology developed in a course of 40 years of testing of blades of various purposes. This methodology can be used when designing a new rotor blade or

DOI: http://dx.doi.org/10.5772/intechopen.1006478

to check the quality of the blade of a known form and shape. These tests include performance testing, fatigue, and structural testing like the one presented in [3–5].

2. Performance testing

The performance testing represents the determination of the traction dynamic characteristics of a propeller, i.e., its thrust force and torque in conjunction with number of revolutions per minute [6].

The test bench for testing the traction-dynamic properties of propellers (1) is built of aluminum profiles (**Figure 1**). This test bench can safely test propellers with diameters of up to 550 mm. The power group comprises one 250 W alternating current motor (2) controlled by a torus precise alternating current regulator (8) with a voltage range of 0–300 V. The torus regulator guarantees that the operational voltage changes continuously. The motor is situated in an aerodynamic formwork constructed of aluminum, which is supported by an aerodynamic pylon (3). A precise scale (7), which is an

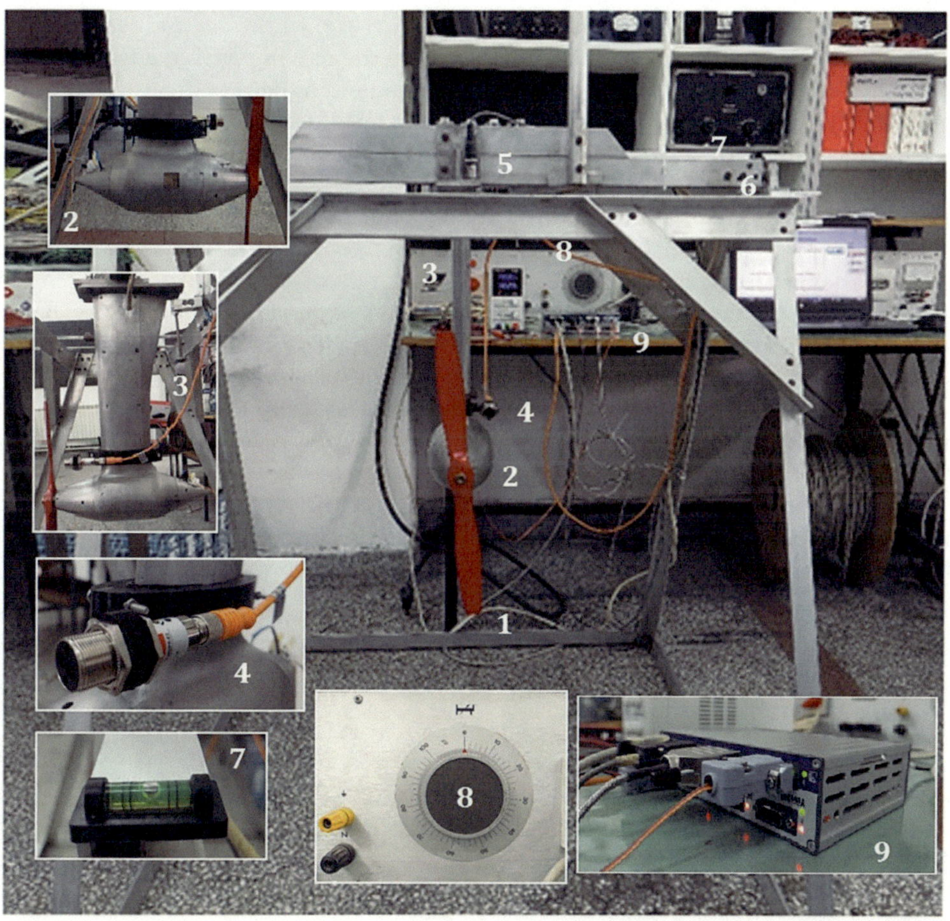

Figure 1.
Experimental setup for traction-dynamic characteristics evaluation.

integrated component, is utilized to finely adjust the propeller's zero balance point, since this test bench can be used for testing of propellers of wide range of diameters, which are of different geometry, hence of different mass. The propeller is mounted on the test bench using a conical holder on the driving shaft that is locked with a self-locking nut.

Propeller's trust is obtained via a consol-type force transducer presented in **Figure 2**. The thrust of the propeller is consequence of its rotation. Number of revolutions per minute of the propeller is obtained with the use of an optical digital encoder LANELO PR18-BC40DPR-E2 (position (4) in **Figure 1**). The encoder is mounted on the aerodynamic support that follows the contour of the pylon and enables the adjustment of the encoder's position in reference to the propeller, if there is the necessity to adjust the measurement response. The propeller's torque is measured via two strain gauges of 120 Ω that are mounted on a special console support that is designed and produced for the purpose of torque measurement. Strain gauge that is used for thrust force measurement is connected to the acquisition device with a Wheatstone full-bridge, while the strain gauges used for torque measurement are connected via Wheatstone half-bridge.

Any propeller imbalance during the test is noticed by the vibration detection system. It consists of two accelerometers (**Figure 3**) that are mounted on different parts of the test bench. One of the accelerometers (Monitran MTN/7100-50, position (6) in **Figure 1**) is placed on a fixed part of the structure, while the other one (HBM 1-B12/200, position (5) in **Figure 1**) is mounted on the moving part of a test bench that is under impact of the thrust force of the propeller. In the case of a resonance of these two sensors, the experiment is automatically terminated.

An eight-channel data acquisition unit QuantumX MX840A is used for measurements that enables 20,000 measurements per second per channel with 24-bit resolution. All 8 A/D converters work synchronously and monitor the transformation of physical quantities into a digital signal in real time. QuantumX contains a LAN and FireWire bus in order to deliver acquired data to the acquisition computer, as well as for expanding the number of measurement points by direct connection to several quantum devices. Before the measurements take place, output of force and torque sensors are set to zero, in order to prevent accumulation error that could emerge on these sensors.

Confirmation of the proposed methodology was carried out by measurements on the propeller shown in **Figure 4**, with a diameter of 420 mm, reference pitch angle at 3/4*D*, and Clark-Y airfoil.

Measurements confirm that the thrust is a square function of the RPM, as is predicted by theory. With increase in RPM the thrust is increased, since the propeller blades rotate faster, thus increasing the air mass flow rate flowing against them. The

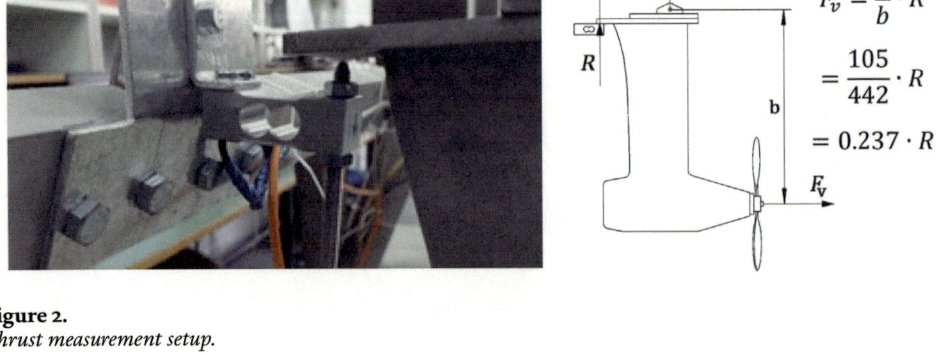

Figure 2.
Thrust measurement setup.

Figure 3.
Accelerometers (HBM and Monitran).

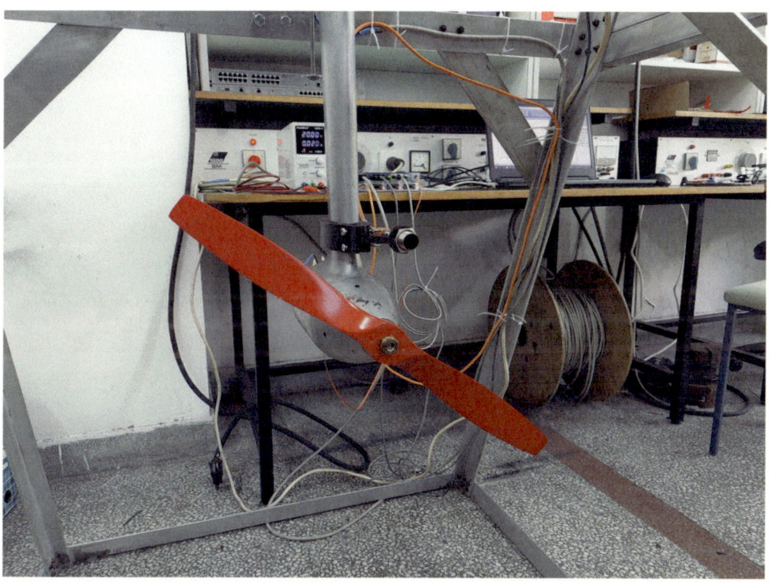

Figure 4.
Propeller ready for testing.

consequence of this is the increase of the forward load applied to the propeller that acts as a moving force forward. Due to the construction of the test bench and the position of the load cell, this motion is restricted by the load cell itself. Hence, the load cell is 'perceiving' the thrust acting on the propeller.

Figure 5 on the right illustrates how the resistance torque varies with RPM. There is a slightly higher variability in the measured values, with a correlation coefficient

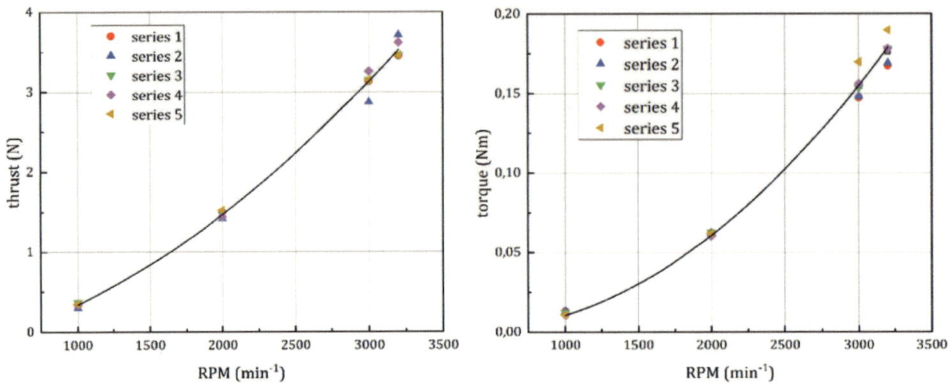

Figure 5.
Thrust force and torque for different values of RPM.

of 0.96. The primary discrepancy in the results occurs at higher RPMs. This variation in torque values is likely due to insufficient stiffness in the console support where the measuring tapes are positioned, leading to increased vibration in the test bench. Consequently, these findings further indicate the quality of the tested propeller.

Figure 6 presents the time dependencies of the signals from the traction force sensor and the resistive torque sensor across all RPMs and measurement series. A clear

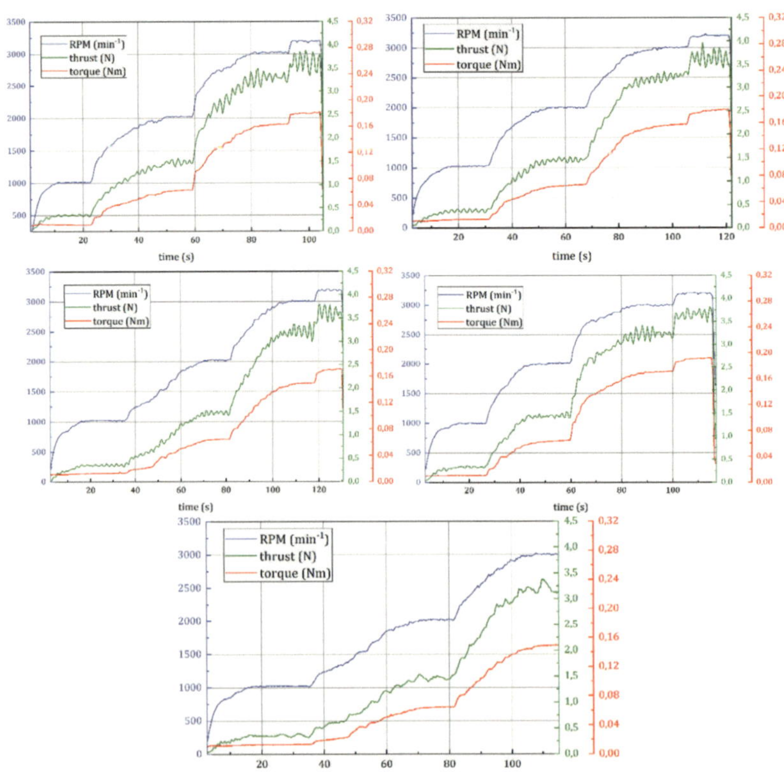

Figure 6.
Time evolution of RPM, thrust, and torque.

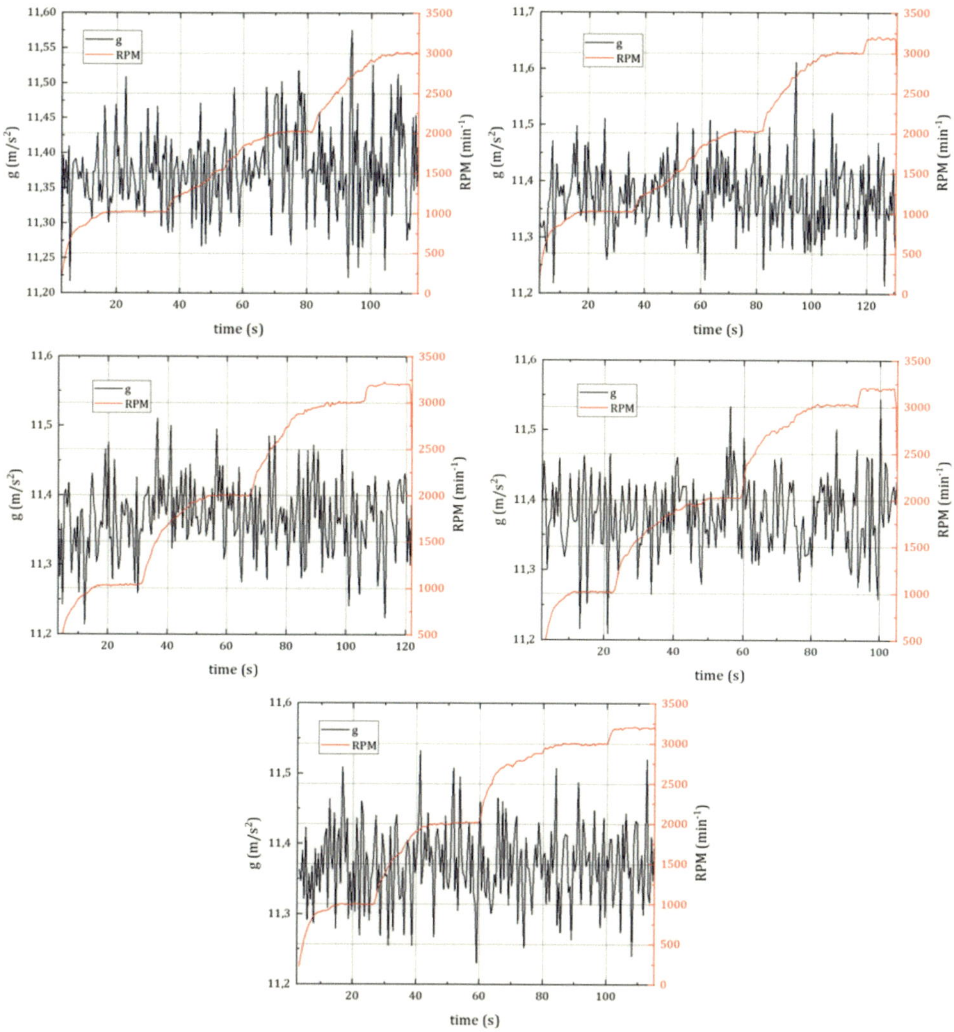

Figure 7.
Signal from accelerometer with the change in RPM.

trend shows that both traction force and torque values track changes in the number of revolutions. Additionally, there is instability in the traction force sensor signal at RPMs exceeding 2000 min^{-1}, which is attributed to increased vibrations on the test bench caused by the imperfections in the tested propeller.

Figure 7 displays the comparative time dependence diagrams of the signals from the vibration transmitter and the optical encoder used to measure the propeller's RPM. In all measurement series, increased vibrations are evident at 2000 min^{-1}, indicating asymmetry in the tested propeller.

3. Mechanical testing

The mechanical testing implies rigidity (flexural and torsional) testing, fatigue testing, and bump test. These series of testing are performed on a single test bench that has two parts. In one part, we can perform rigidity testing, while the other part serves for fatigue and bump testing.

3.1 Rigidity tests

The program for rigidity testing of the composite rotor blade is defined by standard IEC 61400-2 [7]. The purpose of this testing is to define the rigidity of the rotor blade and to determine the maximum force or coupling of forces that cause the failure of the rotor blade as well as the spot of the failure.

The holder for the blade that is meant to be subjected to rigidity testing constitutes of extremely rigid specious grid made of steel C and L profiles that are mutually connected by bolt links. The link between the rotor blade and the holder is accomplished through 30 mm board and by bolts M8 of 8.8 quality. The rotor blade is mounted on the structure at the angle of attack. This is presented on the left of **Figure 8**.

There are two systems for loading in the part of the test bench intended for rigidity testing. The difference is in the nature of the load. When the flexural rigidity testing is performed, we load the blade with the force, while for torsional rigidity it is necessary to load the blade with coupling of forces.

Introducing force is performed by means of specially made system that consists of supporting structure, pulley, reductor unit, and engine with frequent control. The maximum force possible that can be achieved with this system is 2500 daN (**Figure 9**).

The rotor blade is loaded in six sections by lyres and measured from the place of the blade clamp (**Figure 10**, **Table 1**).

The introduction of coupling of forces on the blade, in order to get torsional rigidity, is performed in the way presented in **Figure 11**. In the relevant cross-section of the blade, the size of the required torsional moment is defined, which would cause a twisting angle of one radian in relation to the cross-section in the root part of the blade. A coupling of forces with small magnitudes of the coupling moment is introduced to the blade via the lyre in order to remain in the domain of linearity. The weights of a known mass through the system of pulleys and a lyre that is positioned on the blade are acting on the opposing sides of the blade, forming a coupling of forces.

Figure 8.
Acceptance and mounting of the blades for rigidity testing (left) and fatigue testing (right).

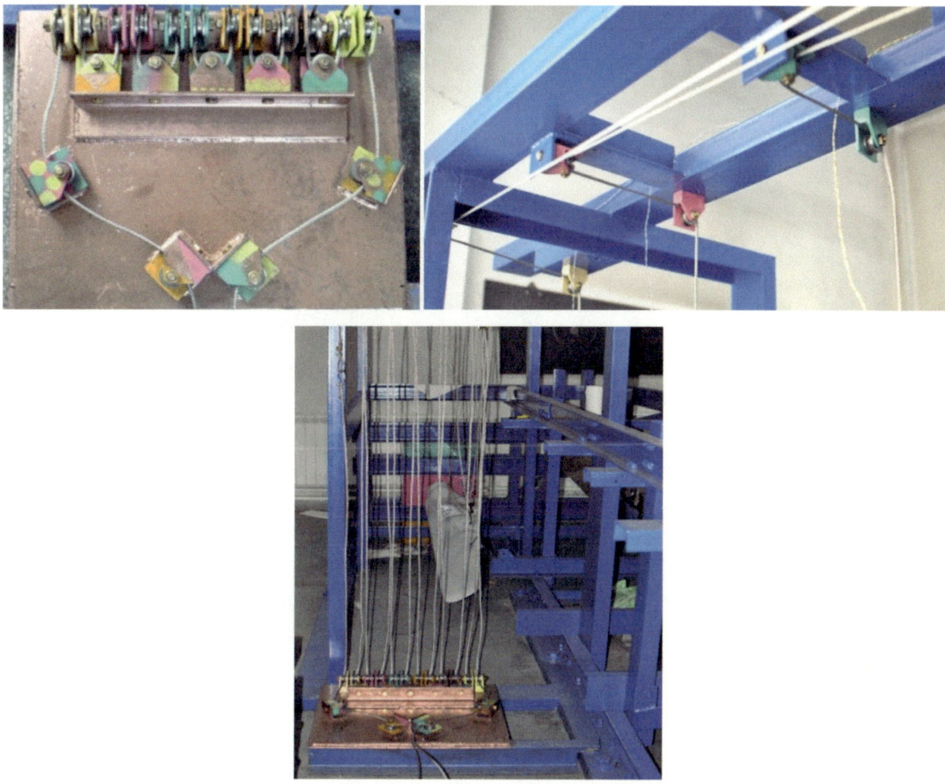

Figure 9.
Constriction for force distribution.

Figure 10.
Lyres for the force introduction.

Section	1	2	3	4	5	6
Distance (mm)	60.0	102.5	154.0	182.5	206.0	236.5

Table 1.
Positions of the six lyres for force introduction, measured from the place of the blade clamp.

Figure 11.
Coupling of forces acting on a blade positioned on the test bench.

The acquisition chain in the case of flexural rigidity testing is consisted of a central acquisition unit, strain gauges, and displacement measuring device. The central acquisition unit HBM SPIDER 8 represents a multifunctional model of receiving analogue and digital signals with parallel tracing of the flow of input units, by means of integrated microcomputer, using higher-level system to relieve the acquisition route and to provide the flow of signals from certain 'smart sensors' directly to the control unit. SPIDER 8 is a multichannel acquisition unit designed for dynamic parallel measuring. Thanks to integration with personal computer as a higher-level system, the process of measuring is remarkably simple, and the total acquisition system is compact and of small dimensions (**Figure 12**).

The SPIDER acquisition unit offers 9600 measurements per second for each channel, with a resolution of 16 bits. All eight A/D converters operate simultaneously, converting physical values into digital signals in real time. For this experiment, two modules are utilized, providing a total of 16 measurement values.

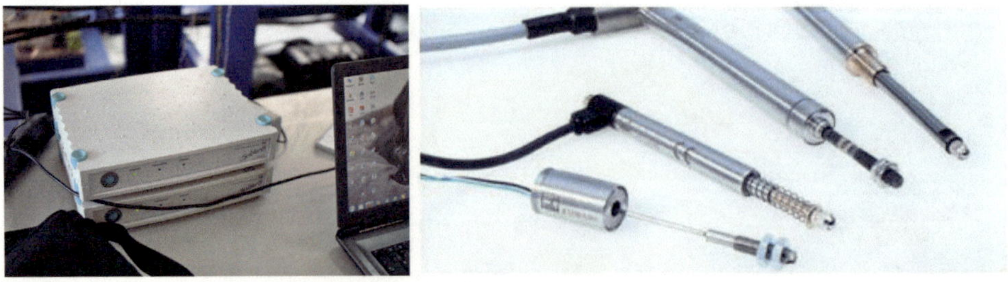

Figure 12.
Two modules of SPIDER 8 and inductive displacement sensor LVDT—HBM.

The movement of a selected spot on the rotor blade is measured using a standard inductive displacement device, the LVDT model from HBM, with an accuracy class of ±0.2% and a measurement range of ±1 to ±500 mm (see **Figure 12**).

Six specially designed force sensors are used for measuring force. Each sensor consists of four strain gauges attached to its body as shown in **Figure 13**. The two strain gauges oriented in the axial direction measure the deformation of the sensor body under load, while the other two gauges, positioned crosswise, compensate for bending deformations. This setup ensures that the measured value is proportional only to the axial load.

Each force sensor (**Figure 14**) is connected to a Wheatstone full measuring bridge. The bridge is powered by a direct input voltage V1, while the output voltage V0 varies based on the resistance changes in the strain gauges R1, R2, R3, and R4 due to deformation. This output voltage is proportional to the axial force applied to the sensor. The arrangement of the strain gauges in the measuring bridge effectively compensates for the effects of bending moments on the output voltage.

The acquisition system for the torsional rigidity testing presented here is not so contemporary. The displacements of the measuring points in a predetermined layout were measured with comparators (**Figure 15**). The value of the torsion angle along its span is determined by means of the measured deflection values on the blade's leading and trailing edges.

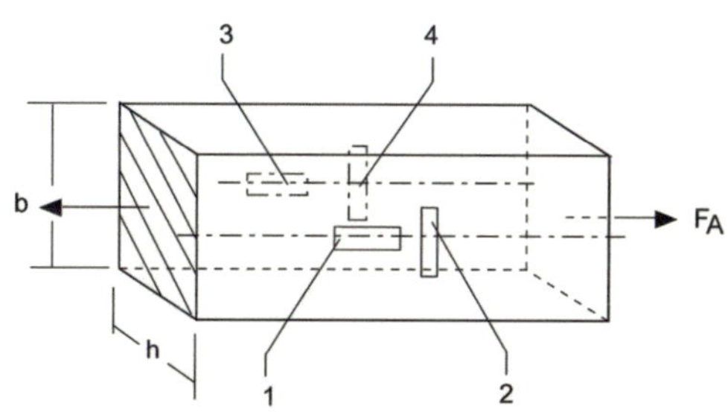

Figure 13.
Schematic of force sensor.

Figure 14.
Force sensor on the blade and positioning on the blade.

Figure 15.
Displacement measurement via comparators in a torsional rigidity testing.

The determination of flexural rigidity is demonstrated on an example of the Scirocco rotor blade—a 5.6 m diameter rotor (**Figure 16**). The rotor blade mass is 11.9 kg. Design and calculations have been made according to IEC 61400 design rules [7, 8].

The rotor blade failure is marked with the resulting force of 490.76 daN under maximum deflection of 27.999 mm. After the testing is completed, some separations are recorded, as presented in **Figures 17** and **18**.

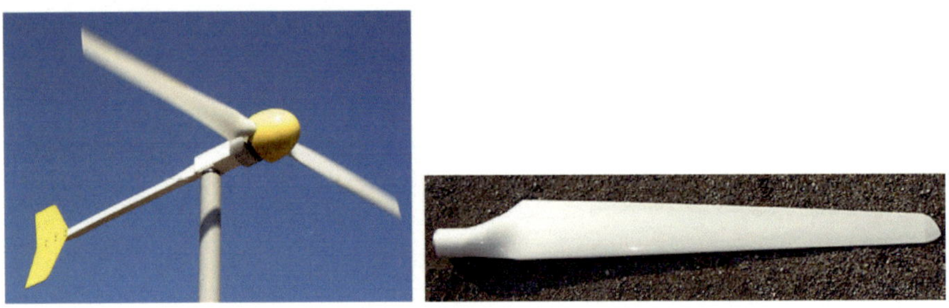

Figure 16.
Wind turbine Eoltec Scirocco and blade W55RBVS.

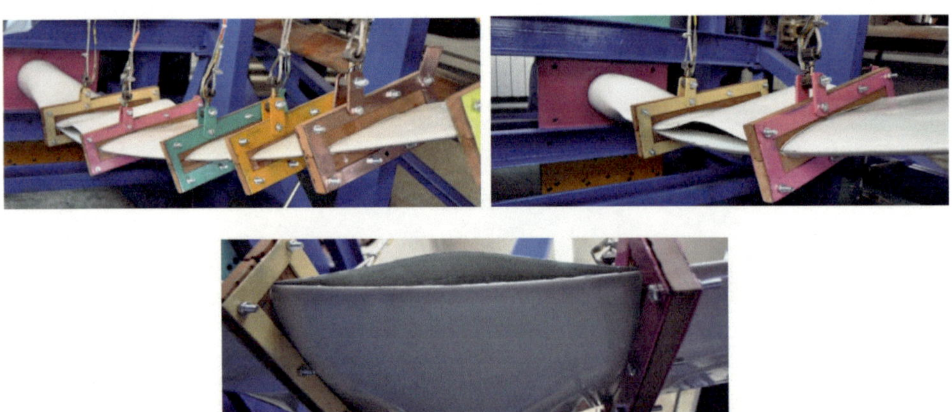

Figure 17.
Position of the blade fracture.

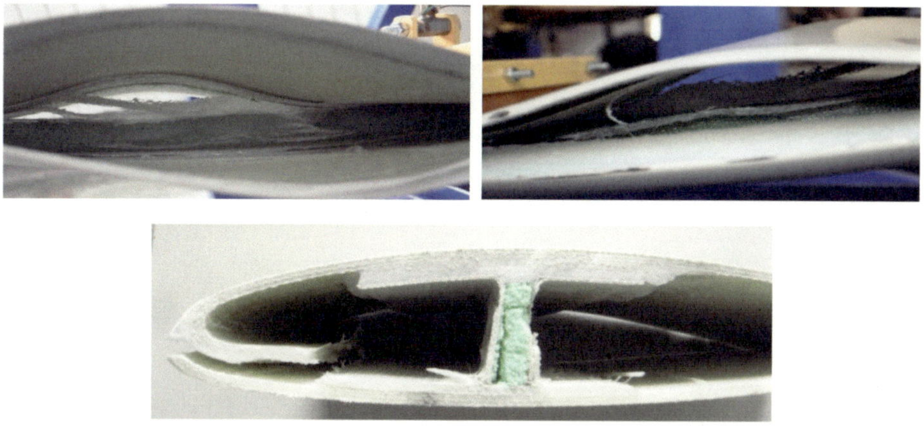

Figure 18.
Layering blades at the fracture.

Figure 19 shows the load applied to each sensor and the cumulative load on the rotor blade that resulted in its fracture. This represents the maximum load that the rotor blade can endure.

Next, we present the results of torsional rigidity testing for the tail rotor blade of the MI-8 helicopter, which is primarily a metal structure in its basic version. The composite blade of the MI-8 tail rotor serves as a long-term test sample at the Aviation Institute of the Faculty of Mechanical Engineering in Belgrade [9, 10]. It has successfully completed all testing stages outlined by this methodology. **Figure 20** shows the cross-sections where the torsional rigidity will be experimentally determined.

The measured values of torsional rigidity in four cross-sections of a rotor blade are presented in **Figure 21**. It is obvious the influence of the root of the blade on its torsional rigidity in cross-section 1. This test is performed in an elastic area of the blade's material. There is a possibility of determining the coupling of forces value that will lead to the failure of the rotor blade, which was not done in the presented case.

3.2 Fatigue test

On the right side of **Figure 8**, the setup for the blade prepared for bump or fatigue testing is shown. The connection is made using two plates and a double spindle with a trapezoidal thread. The angle of attachment of the blade is adjustable, with the front plate marked for the angle of attack, and the front nut secures the blade in place.

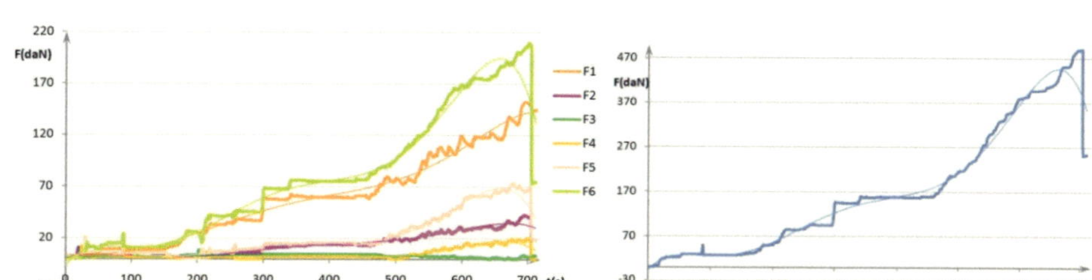

Figure 19.
Load distribution of each section in function of time and resulting force distribution in function of time.

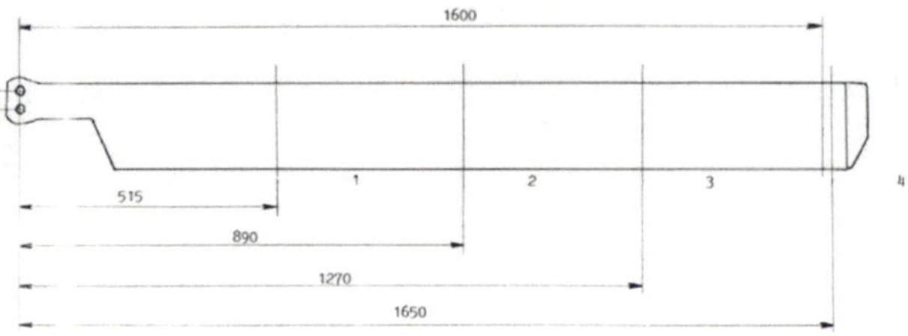

Figure 20.
Cross-sections on the rotor blade for torsional rigidity testing.

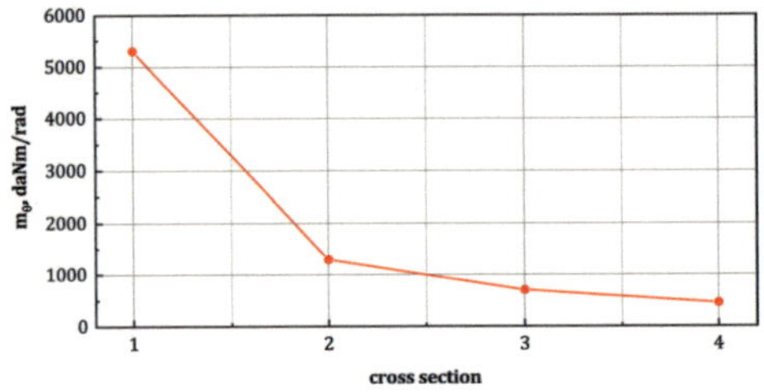

Figure 21.
Experimental results of torsional rigidity in four cross-sections.

The connection is fully secured against unscrewing with the back plate and nut. Additionally, a lever and pulley provide extra security against rotation in the clamping plane.

We will present the results of fatigue testing for the tail blade of the MI-8 helicopter, made from composite material at our Institute. The fatigue test bench features a hydraulic system for applying axial load to the blade, which includes a hydraulic cylinder, pressure gauges, an oil distribution system, and a pump with a drive motor. This test bench is a unique solution designed and implemented in the wind tunnel laboratories of the Faculty of Mechanical Engineering at the University of Belgrade. A schematic of the test bench is shown in **Figure 22**.

The static load is simulated by a centrifugal force of 11,350 daN, which results from the inertial characteristics of the blade at a nominal speed of 1100 RPM, and is applied using a hydraulic cylinder. The registration of this centrifugal force is facilitated by a strain gauge, a measuring bridge, and a suitable acquisition system. The strain gauges are attached to the transducer and configured in a full Wheatstone bridge (**Figure 23**) to enhance sensitivity and provide temperature compensation. Calibration of the measuring system was performed up to 20,000 daN using an

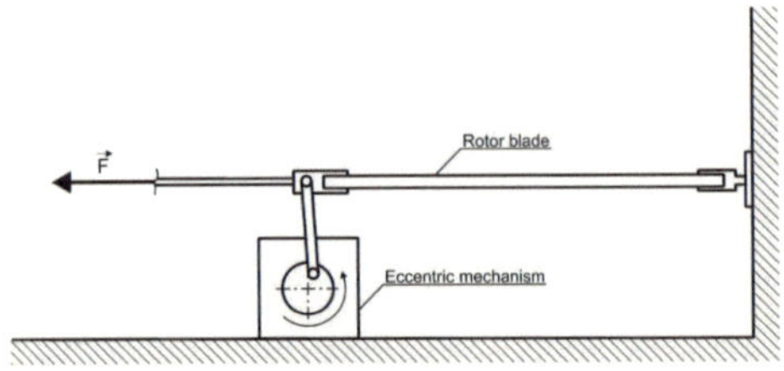

Figure 22.
Schematic of the test rig for fatigue testing.

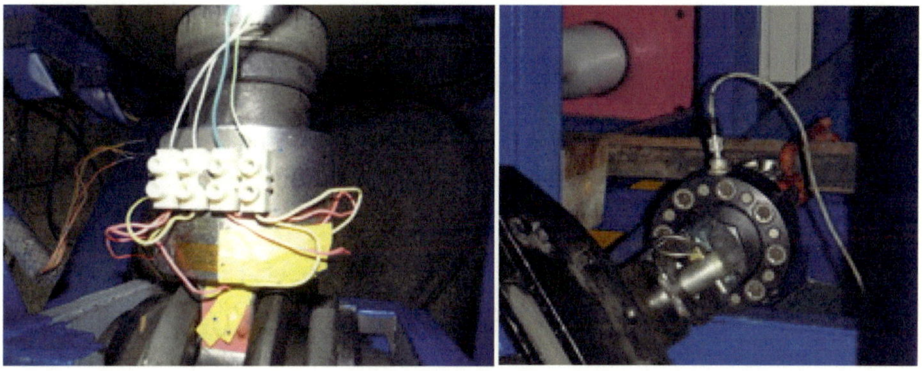

Figure 23.
Integrated strain gauges (left) and reference force transducer (right).

HBM-calibrated inductive transducer, which has satisfactory measurement uncertainty (**Figure 23**). The SPIDER 8 served as the central acquisition unit for this testing, with its characteristics described previously.

The driving force simulating the alternating load is 500 daN, determined primarily by the aerodynamic characteristics of the blade, and acts normal to the plane of rotation. The registration of the excitation force is carried out in the same manner as the centrifugal force measurement, using strain gauges connected in a full Wheatstone bridge and an acquisition system. The strain gauges are attached to the excitation lever that connects the eccentric to the blade. The value of the excitation force is adjusted using an eccentric with variable eccentricity. Calibration of the strain gauges for an excitation force of 600 daN is performed on a special scale with a lever transmission ratio of 1:10. The frequency of the excitation force changes at 6.5 Hz (390 RPM), and the test is conducted at an angle of attack of 18°.

With the simultaneous application of these forces during the fatigue test, a total of 1500,000 cycles will be conducted at a frequency of 6.5 Hz. Following this, the relaxation phase will involve 1500 relaxations of the static centrifugal force, varying from 0 to 11,350 daN, in accordance with the guidelines outlined in the relevant document [11].

Due to the nature of the simulated load, continuous monitoring and control were conducted through the acquisition system's displays. The frequency of the excitation force was checked and adjusted at the beginning and end of each testing period. Before testing commenced, a thorough inspection of the tail rotor blade was performed. Additionally, careful monitoring of the blade's behavior took place throughout the test. No signs of delamination in the composite structure were observed during this time. Following the test program, a detailed inspection of the blade revealed no traces of damage, degradation, or changes in shape or structure, confirming that no delamination had occurred.

The fatigue test of the rotor blade demonstrated that it is capable of performing all its essential functions on the helicopter. This composite material rotor blade successfully endured the entire test program under all relevant loads, confirming the quality of the manufacturing technology used for the tail blade with composite materials.

3.3 Bump test

The bump test is conducted to determine the natural frequencies (frequency characteristics) of the examined blade. This test involves impulse excitation and measuring the blade's response. The excitation is achieved by delivering a short impact with a hammer to the blade. For expected natural frequencies below 5000 Hz, a rubber hammer is recommended, while a plastic or wooden hammer is suitable for higher frequencies; thus, a rubber hammer was used in this case. It is important to ensure that the impact is brief enough to avoid damping the free oscillations that occur afterward. The free blade, mounted as it would be on a helicopter rotor, oscillates at all its natural frequencies following excitation. By recording displacement, velocity, or acceleration in the time domain and transforming these measurements into the frequency domain, the natural frequencies of the blade can be obtained. For this test, HBM accelerometers were used, placed at one-third intervals along the length of the blade (**Figure 24**), in conjunction with the SPIDER 8 acquisition system previously discussed in this chapter.

The connection between the time and frequency domains is illustrated in **Figure 25**. The complex signal in the time domain consists of two sinusoids with

Figure 24.
Position of the accelerometers.

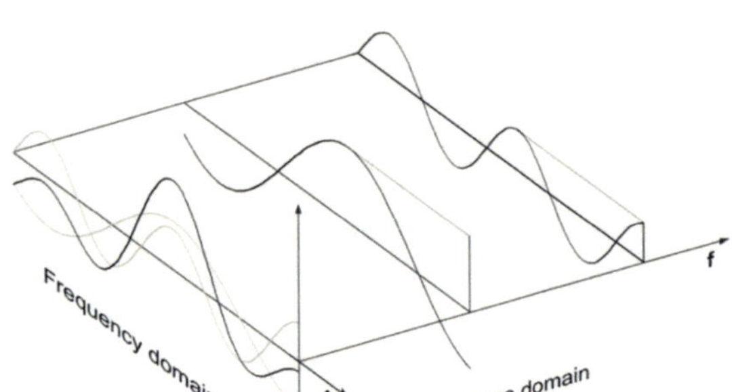

Figure 25.
Connection between the time and frequency domains.

different frequencies. By plotting the amplitude values of these sinusoids on the frequency axis, the representation of the signal in the frequency domain is obtained.

In our case, frequency (spectral) analysis is performed using Fourier analysis, which is based on the assumption that every periodic signal can be divided into a certain number of sinusoids, the sum of which is actually the signal we are observing. In contrast to continuous signals for which integral Fourier transformation (IFT) is used for analysis, discrete Fourier transformation (DFT) is used for discrete signal analysis. Fast Fourier transformation (FFT) is an algorithm that allows a computer to quickly perform a discrete Fourier transformation. Since the continuous signal coming from the transmitter is converted in the A/D converter into a discrete signal which is then analyzed in the computer, DFT was used for frequency analysis, that is, the algorithm for its calculation FFT. **Figure 26** presents the results of bump tests at different blade loads.

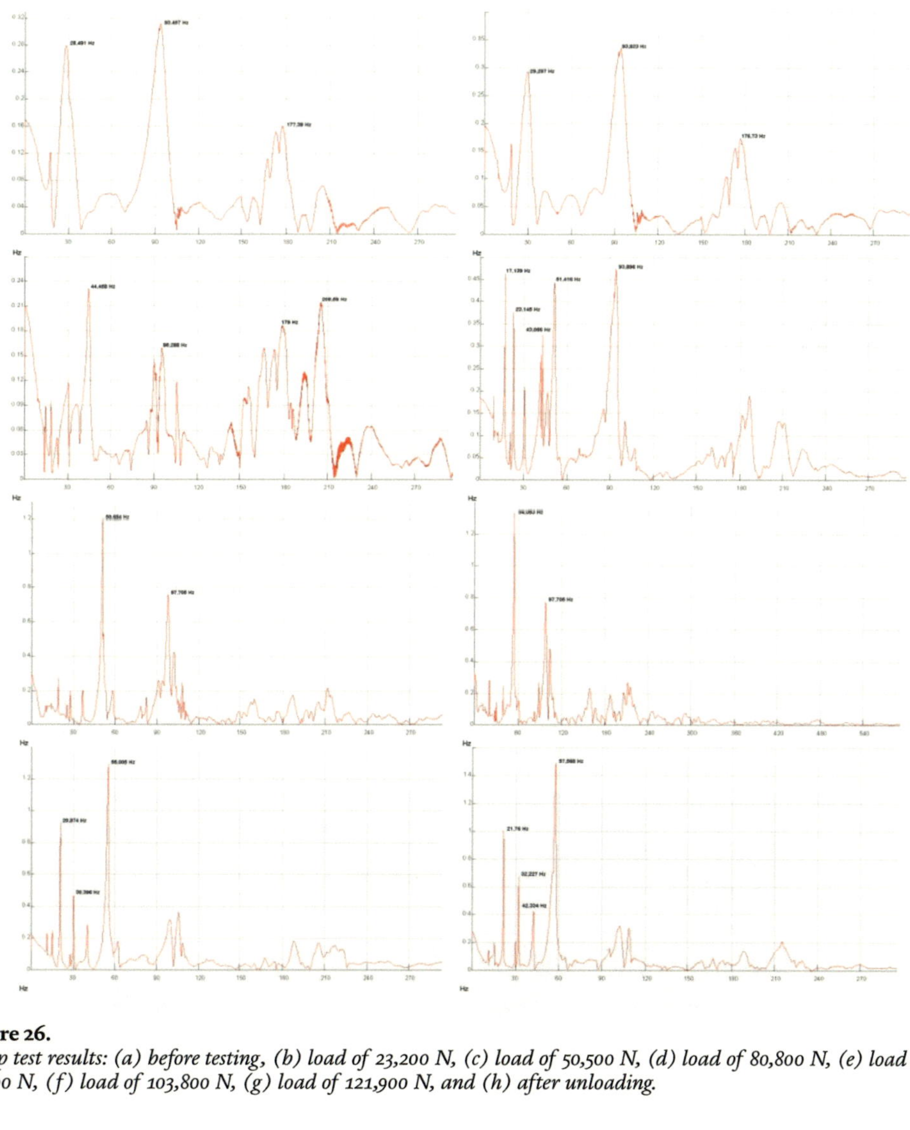

Figure 26.
Bump test results: (a) before testing, (b) load of 23,200 N, (c) load of 50,500 N, (d) load of 80,800 N, (e) load of 98,800 N, (f) load of 103,800 N, (g) load of 121,900 N, and (h) after unloading.

4. Structural testing by AFM/MFM mode

This research utilizes atomic and magnetic force microscopy to examine differences in surface topography and magnetic properties between sections of a rotor blade subjected to varying mechanical loads during rigidity testing. These findings could offer new insights into the material structure of rotor blades and potentially enhance their mechanical performance. The impellers were subjected to critical loads until failure, and sections from both critically and sub-critically loaded areas of the rotor blade are analyzed using magnetic force microscopy (MFM). MFM enables the characterization of surface and near-surface internal structures of samples, which is employed in this study to analyze structural changes in materials under different loads. The study presents results and compares the microstructures of two segments of the wind turbine blade. Additionally, atomic force microscopy (AFM) was employed to observe the surface morphology of the composite samples [12].

In this study, the scanning probe microscope utilized is the SPM-5200 by JEOL, Japan. Typically, deflection is measured by directing a laser spot onto the top surface of the cantilever, which is then reflected into an array of photodiodes. TappingMode AFM represents a more recent advancement where the imaging probe undergoes vertical oscillation at or near the resonant frequency of the cantilever (see **Figure 27**). Electromechanical feedback is employed to maintain constant oscillation amplitude during scanning. The resulting image is generated by mapping the vertical distance maintained by the scanner at each lateral data point to preserve the constant oscillation amplitude (see **Figure 28**). A significant advantage of TappingMode is its ability to eliminate lateral shear forces inherent in contact mode, which can potentially damage the structure being imaged, particularly on delicate specimens.

Magnetic force microscopy (MFM) is a derivative imaging mode based on TappingMode AFM, which maps the magnetic force gradient above the sample surface (refer to **Figure 29**). Unlike traditional AFM, MFM employs a probe tip coated with a thin layer of ferromagnetic material. This specialized tip reacts to magnetic domains present on the sample surface.

To obtain an image of the sample, MFM utilizes a two-pass technique. During the initial scan, the topography of the sample is mapped out. In the subsequent scan, the tip-sample distance is increased, and the biased tip is scanned along the topographical features obtained from the first scan. This second scan ensures that the tip maintains a constant distance from the sample surface, corresponding to the line of constant van der Waals force.

During the second scan in magnetic force microscopy (MFM), the tip follows the topographical line while maintaining a constant van der Waals force. This ensures that changes in the signal primarily reflect variations in magnetic force. The evaluation of

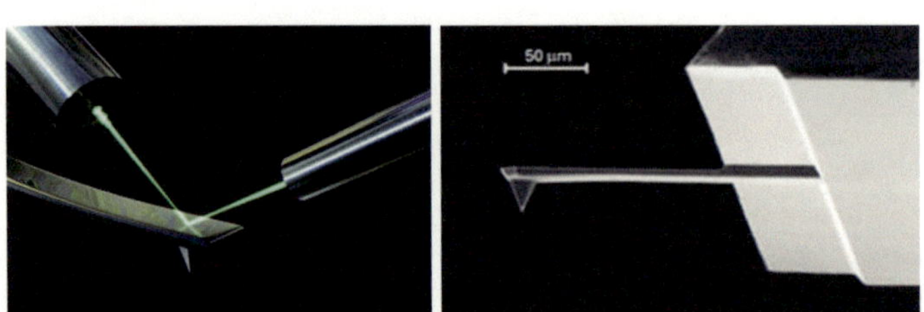

Figure 27.
Cantilever with a sharp tip (probe).

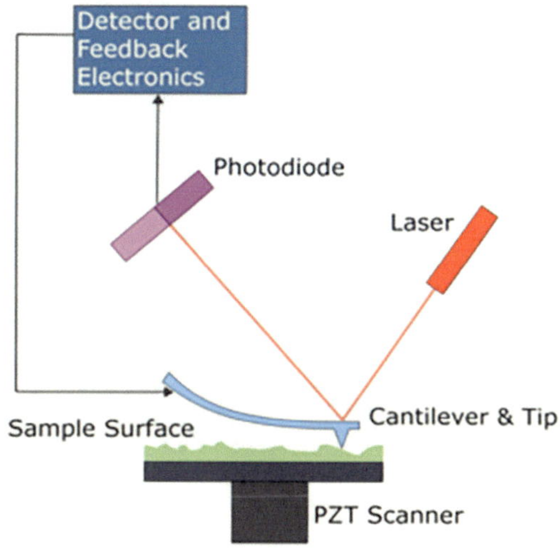

Figure 28.
Schematic of a laser measurement.

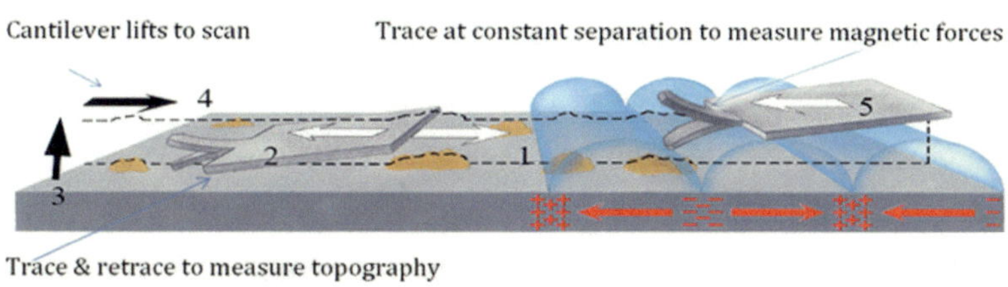

Figure 29.
LiftMode for topography and magnetic force gradient measurement.

magnetic properties relies on the magnetic force gradient image, which qualitatively depicts the distribution of magnetic fields across the scanned area, indicating local variations in magnetic properties.

The cantilever used in this study is manufactured by MikroMasch (Estonia) under the trade name NCS18 Co-Cr. The MFM probe features a silicon etched tip with a conical shape, coated with layers of Co and Cr. As a result of the coating, the tip has a radius of approximately 90 nm. The full cone angle of the tip is 40°.

It is anticipated that the less loaded portion of the material (Sample 1, depicted in **Figures 30** and **31**) will exhibit lower surface roughness. This is because the layers of silicone fibers held by epoxy are expected to be more uniformly aligned under lower stress conditions. AFM analysis reveals that Sample 1 has an average surface roughness (Ra) approximately 200 nm smaller than Sample 2, which experienced higher loads during wind turbine testing. This suggests that the inner layers of Sample 2 may have undergone deformation, potentially resulting in misalignment compared to their original state.

From the magnetic characteristic figures, it is evident that Sample 1 displays a relatively uniform distribution of magnetic properties, whereas Sample 2 exhibits 'black dots'. These 'black dots' signify abrupt changes in sample magnetization, indicative of variations in magnetic field gradients. The presence of these features on the figure of Sample 2, which was subjected to higher loads (as shown in **Figure 32**), suggests

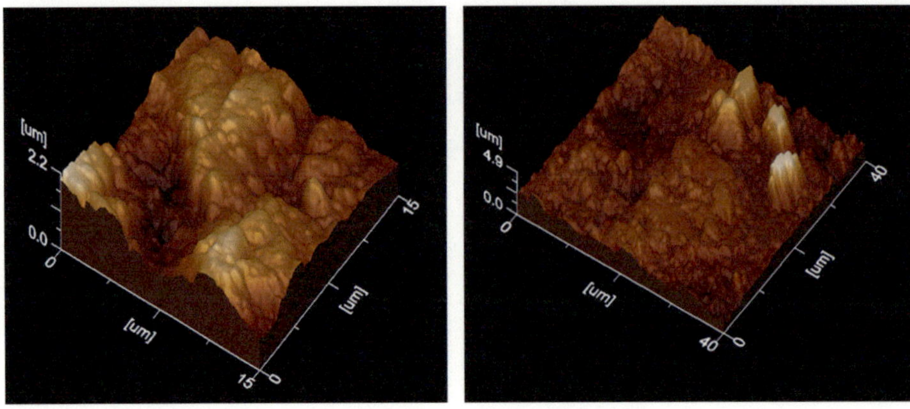

Figure 30.
Surface roughness for less loaded (left) and more loaded (right) sample.

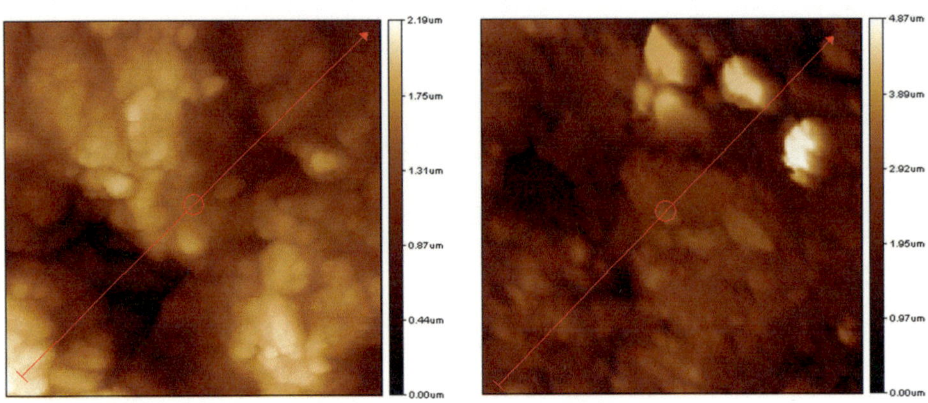

Figure 31.
Detail of surface roughness for less loaded (left) and more loaded (right) sample.

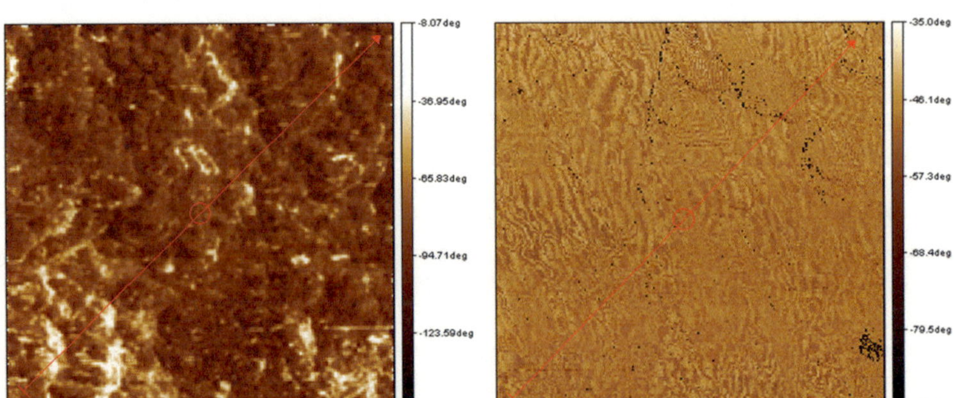

Figure 32.
Magnetic characteristics of the material: left—less loaded, right—more loaded.

potential alterations in the material's structure and stratification due to the applied loading conditions.

Research using atomic and magnetic force microscopy indicates that samples taken from areas under higher loads exhibit the expected characteristics of increased surface roughness and height differences. Additionally, the presence of 'black dots' or 'magnetic holes' suggests internal stratification within the composite material. Overall, these findings from the experiment confirm that the tested rotor blade demonstrates very high rigidity.

5. Conclusions

In this chapter, methodology for testing a rotor blades made of composite material is presented. It involves performance and mechanical testing. These tests are needed, either in the course of design or at the end of production of a rotor blade. Presented example for performance testing proves that the test bench for performance testing can be used both for testing the traction-dynamic characteristics and for propellers' balancing. Test bench for rigidity testing can be used in the area of elastic deformations of the rotor blades, hence determining the flexural/torsional rigidity of the blade on the test. However, it can be used to experimentally determine the maximum force/coupling of forces that the blade on the test can withstand before the failure. In this chapter, we have presented determination of torsional rigidity (elastic area of deformation) and maximum force that led to destruction of the rotor blade. Test bench for fatigue testing is also presented. Bump test reveals at what frequencies the rotor blade should not be operating on. The rigidity tests should be complemented with atomic and magnetic force microscopy to discern variations in surface topography and magnetic properties among different sections of the rotor blade subjected to varying mechanical loads.

Presented methodology can be applied for rotor blades of any size, depending on the available hardware on the site. Of course, for extremely large structures, one can apply the similarity theory and dimensional analysis. The acquisition system is simple and the same, independently on the size of the test bench.

Author details

Goran Vorotović[1*], Jela Vorotović[1], Nebojša Petrović[1], Nikola Davidović[1], Đorđe Novković[1] and Pavle Petrović[2]

1 Faculty of Mechanical Engineering, University of Belgrade, Belgrade, Serbia

2 Civil Aviation Directorate of the Republic of Serbia, Belgrade, Serbia

*Address all correspondence to: gvorotovic@mas.bg.ac.rs

References

[1] Armstrong KB, Barrett RT. Care and Repair of Advanced Composites. 1st ed. Warrendale, PA, USA: SAE International; 1998

[2] Hoskin BC, Baker AA. Composite materials for aircraft structures. In: AIAA Education Series. 1st ed. New York: AIAA Inc.; 1986

[3] Adams DO, Kearney HL. Full-scale fatigue testing of advanced fiber composite components. Journal of the American Helicopter Society. 1986;**31**:66-72. DOI: 10.4050/JAHS.31.66

[4] Curtis PT. The fatigue behaviour of fibrous composite materials. The Journal of Strain Analysis for Engineering Design. 1989;**24**(4):235-244. DOI: 10.1243/03093247V244235

[5] Mitrović Č, Petrović N, Bengin A, Bekrić D, Dragović V, Simonović A, et al. Structural testing of small wind turbine blade up to failure. In: International Conference on Innovative Technologies. Conference IN-TECH 2011, Bratislava, CTU – Czech Technical University in Prague. 2011

[6] Vorotović G, Burazer J, Bengin A, Mitrović Č, Januzović M, Petrović N, et al. A case study of a methodological approach to the verification of UAV propeller performance. Strojniški Vestnik—Journal of Mechanical Engineering. 2023;**69**(5-6):199-207. DOI: 10.5545/sv-jme.2022.432

[7] International Standard IEC 61400- 2. Wind Turbines – Part 2: Design Requirements for Small Wind Turbines. Switzerland: International Electrotechnical Commission; 2019

[8] International Standard IEC 61400-1. Wind Turbines - Part 1: Design Requirements. Switzerland: International Electrotechnical Commission; 2019

[9] Rasuo B. Design, fabrication and testing of the helicopter tail rotor blade from composite laminated materials. In: 13th International Conference on Composite Materials, ICCM-13. Beijing, China: International Committee of Composite Materials; 2001

[10] Rasuo B. Testing of the Survivability of an Aeronautical Constructions from Composite Materials, SAE/AIAA 1999- 01-5597. In: World Aviation Congress. San Francisco, California: SAE International; 1999

[11] Rasuo B. Helicopter tail rotor blade from composite materials: An Experience. SAE International Journal of Aerospace. 2011;**4**(2):828-838. DOI: 10.4271/2011-01-2545

[12] Yahya N, Kashif M, Daud H, Mohd Zaid H, Shafie A, Nasir N, et al. Fabrication and characterization of Y3.0-XLaXFe5O12 PVA composite as EM waves detector. International Journal of Basic & Applied Sciences IJBAS—IJENS. 2009;**9**(9):131-134

Chapter 3

Advancements in Aerospace Engineering: Exploring Emerging Technologies for Advanced Aerospace Efficiency

Abhay Dhasmana and Sachin Srivastava

Abstract

Currently, conventional aerospace technologies face numerous technological challenges that include environmental pollution, budget constraints, safety standards, etc. The rapid development of technologies has made it possible to address and solve these limitations. This chapter discussed the latest technological innovations from aerospace materials to alternate fuel and propulsion systems all to enhance the efficiency of flights. Some cutting-edge technologies like Autonomous systems and quantum computing which can open new opportunities in the aerospace industry are also discussed in this chapter. Advanced propulsion systems such as electric propulsion can solve the issues of environmental pollution by reducing the dependency of the current aerospace industry on fossil fuels. Various emerging concepts such as Industry 4.0 and Industry 5.0 can also be efficiently integrated into the aerospace sector to not only boost performance while flying but also improve the aerospace part assembly and structural analysis. By addressing these latest research and technologies, this chapter aims to leverage and employ these latest emerging technologies for the benefit of the aerospace sector.

Keywords: aerospace engineering, autonomous systems, quantum computing, Industry 4.0, Industry 5.0, propulsion systems

1. Introduction

The aerospace industry is currently undergoing dynamic changes due to the integration of innovative technologies and advanced materials. This chapter explores key advancements shaping the future of aerospace, focusing on advanced aerospace materials, alternative fuel and propulsion systems, autonomous systems, quantum computing, and the integration of Industry 4.0 and Industry 5.0. Each of these areas represents a significant development, promising to enhance performance, efficiency, and sustainability in aerospace applications.

Advanced materials are essential in developing lighter, stronger, and more efficient aerospace structures. Lightweight composite materials, such as carbon fiber-reinforced polymers, are reducing aircraft weight and improving fuel efficiency [1]. High-temperature alloys, including nickel-based superalloys, are essential for modern propulsion systems that operate under extreme conditions [2]. Smart materials that can adapt to changing conditions offer the potential for enhanced performance and safety in aerospace applications [3].

The propulsion systems are evolving with advancements in electric propulsion, which promises reduced environmental impact and increased efficiency [4]. Hydrogen fuel cells are emerging as a viable alternative, offering zero-emission energy sources that can significantly reduce the aerospace industry's carbon footprint [5]. Additionally, biofuels are being explored as sustainable alternatives to traditional aviation fuels, with research focusing on their integration into existing engines and infrastructure [6].

Unmanned Aerial Vehicles (UAVs) have revolutionized aerospace operations, providing capabilities for surveillance, reconnaissance, and even cargo delivery [7]. Autonomous navigation and control systems enhance the reliability and efficiency of UAVs, using advanced algorithms to perform complex tasks without human intervention [8].

Quantum computing holds transformative potential for aerospace by solving complex problems related to optimization, simulation, and data analysis. Quantum algorithms, such as those for simulated quantum computation of molecular energies, can significantly enhance the development of advanced materials and propulsion systems [9]. The utilization of variational hybrid quantum-classical algorithms is pushing the boundaries of what can be achieved in computational chemistry and materials science [10]. The future of quantum computing in aerospace promises not only improvements in computational power but also new capabilities in real-time data analysis and optimization [11].

The integration of Industry 4.0 and Industry 5.0 technologies is revolutionizing aerospace manufacturing and operations. Digital twins and advanced simulation techniques provide real-time insights and predictive capabilities, enhancing the efficiency and accuracy of aerospace design and maintenance [12]. The Internet of Things (IoT) enables detailed monitoring and data collection, leading to more informed decision-making and predictive maintenance [13]. Advanced manufacturing techniques, such as 3D printing, enhance the efficiency of the manufacturing process and enable the fabrication of complex components while minimizing material waste [14]. Industry 5.0, with its focus on human-centric and sustainable approaches, further enhances the flexibility and adaptability of aerospace operations by integrating advanced technologies with human expertise [15].

This chapter offers a thorough summary of the most recent developments in aerospace technology, ranging from innovative manufacturing and computer techniques to innovative materials and propulsion systems. Through an analysis of these domains, we highlight the revolutionary capacity of these technologies to influence the development of aerospace engineering in the future, guaranteeing enhanced performance, efficiency, and sustainability.

2. Advanced aerospace materials

2.1 Lightweight composite materials

Lightweight composite materials are becoming significantly important daily due to their exceptional strength-to-weight ratios, performance, and durability over conventional materials. These materials are manufactured by combining two or more materials with different properties; due to this, the resultant material exhibits superior mechanical properties compared to individual components. These composites are primarily utilized to boost the efficiency, reduce the weight, and improve the overall performance of aerospace structures. Fiber-reinforced polymers (FRPs), metal matrix composites (MMCs), and ceramic matrix composites (CMCs) are the primary types of composites used in the aerospace industry. These lightweight composites provide a high strength-to-weight ratio, which is crucial in aerospace applications as it can reduce the weight, leading to increased payload capacity. The CFRPs (carbon fiber-reinforced polymers) provide high tensile strength while being lighter than metals like aluminum and steel [2].

Composite materials also offer excellent fatigue resistance, which is essential reliability of aerospace components. It is shown that composites can withstand cyclic loading better than traditional materials, reducing the risk of material failure over time. This is especially beneficial for structures such as wings and fuselage structures which experience repetitive stress [16]. These composites also show corrosion resistance, unlike many metals which when exposed to harsh environments can corrode and weaken over time. Composites having polymer matrices exhibit enhanced resistance to environmental degradation. This enhances the lifespan and reduces maintenance costs of aerospace components [17].

Lightweight thermoset composites are also widely used in the aerospace industry due to their high thermal stability and excellent mechanical properties. Thermoset composites, such as epoxy-based CFRPs, are processed with the help of curing, which involves a chemical reaction that hardens the material. This process results in a robust and dimensionally stable composite suitable for high-stress aerospace applications [18]. Composites can be manufactured depending on their application through a variety of processes such as hand lay-up, filament winding, resin transfer molding (RTM), automated fiber placement (AFP), etc. There is still a lot of important research that is ongoing to improve the properties of aerospace composites, such as enhancing their impact resistance and thermal conductivity. Nanocomposites is a field of study that holds the potential to further develop the aerospace industry by creating advanced materials with enhanced strength and multi-functional capabilities (**Figure 1**) [19].

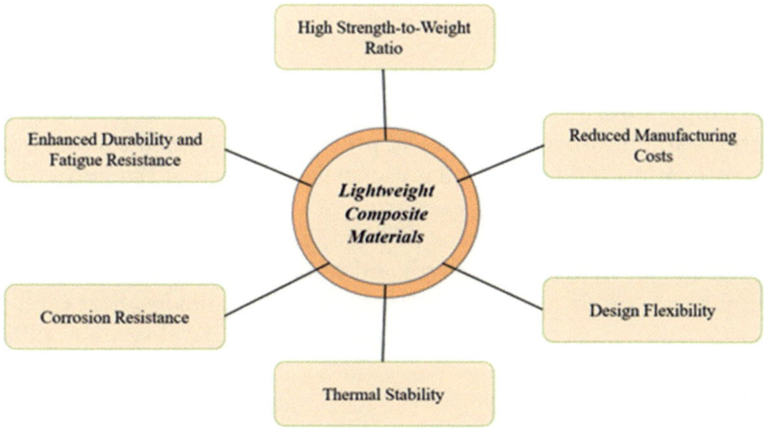

Figure 1.
Benefits of lightweight composite materials.

2.2 High-temperature alloys

High-temperature alloys are essential in the aerospace industry, where materials must withstand extreme conditions without losing their mechanical properties. These alloys offer exceptional performance in high-temperature environments, making them essential for various aerospace applications. Ni-based superalloys and γ-TiAl alloys are examples of these High-temperature alloys.

2.2.1 Ni-based superalloys

Ni-based superalloys are widely used in the aerospace industry due to their high-temperature strength, corrosion resistance, and ability to maintain structural integrity at extreme temperatures. Ni-based superalloys majorly consist of nickel, with significant additions of chromium, cobalt, and other elements that enhance their high-temperature performance. These alloys exhibit excellent creep resistance, which is critical for components exposed to prolonged high temperatures [20]. These superalloys are extensively used in turbine blades, combustion chambers, and other critical components of jet engines, where they contribute to improved efficiency and longevity [21]. Ni-based superalloys having complex geometries and optimized microstructures are manufactured using advanced processing methods such as powder metallurgy and additive manufacturing [22].

2.2.2 γ-TiAl alloys

γ-TiAl alloys are also high-temperature materials that have gained popularity in aerospace applications due to their lightweight and high strength-to-weight ratio. γ-TiAl alloys are majorly composed of titanium and aluminum, offering excellent high-temperature strength, oxidation resistance, and low density compared to traditional superalloys [20]. These alloys are used in low-pressure turbine blades, automotive valves, and other components where weight savings and high-temperature performance are essential [23]. The processing of γ-TiAl alloys involves advanced techniques such as directional solidification and hot isostatic pressing to achieve the desired mechanical properties and microstructure [24].

2.2.3 Rapid solidification processing

Rapid solidification processing (RSP) is a technique used to enhance the properties of high-temperature alloys by rapidly cooling the molten alloy, leading to fine microstructures and improved mechanical properties. The rapid cooling process minimizes segregation and produces a uniform distribution of strengthening phases, which are essential for maintaining performance at elevated temperatures [21]. RSP involves cooling the alloy at rates exceeding 10^5 K/s, resulting in refined microstructures that enhance the material's strength and resistance to high-temperature degradation. Components produced using RSP include turbine blades, heat exchangers, and other critical parts where high strength and thermal stability are required [22].

2.3 Smart materials

Smart materials are advanced materials that can respond dynamically to external stimuli. These materials possess properties that can be significantly altered in a

DOI: http://dx.doi.org/10.5772/intechopen.1006686

controlled environment by external stimuli, such as stress, temperature, moisture, pH, and electric or magnetic fields. Their integration into aerospace structures promises to enhance performance, improve safety, and reduce maintenance costs. The initial exploration of smart materials for aerospace applications can be traced back to the late twentieth century. These early studies focused on materials like shape-memory alloys and piezoelectric materials, which could change shape or generate an electric charge in response to external stimuli. These capabilities were particularly useful for developing adaptive structures and improving aircraft performance and safety [25].

Smart materials have seen significant advancements in their applications within the aerospace industry. These materials are being increasingly used to develop more efficient, safer, and cost-effective aerospace components. Applications range from vibration damping and noise reduction to adaptive wing structures and real-time structural health monitoring systems. These applications improve the operational efficiency of aerospace systems and contribute to extending the lifespan of components and reducing overall maintenance costs [26]. The development of smart materials for aerospace sensing systems is also progressing rapidly. Smart materials, such as piezoelectric and shape-memory alloys, are being used in sensors to monitor the structural health of aircraft in real time. These materials can detect stress, strain, temperature changes, and other factors that may affect the integrity of aerospace structures. The ability of these materials to provide real-time data allows for immediate responses to potential issues, thereby enhancing the safety and reliability of aerospace operations [27].

The new generation of aerospace applications involves more complex and advanced uses of smart materials. Smart materials have the potential to create self-healing structures and components that can autonomously respond to damage. This not only improves the durability and lifespan of aerospace components but also reduces the need for frequent inspections and repairs [28]. Smart materials are essential for the development of active structures in aerospace engineering. Active structures are those structures that can adapt their properties in response to changing environmental conditions. This includes morphing wings that can change their shape for optimal aerodynamics, or fuselage sections that can adjust their stiffness for better load distribution. The integration of smart materials into these active structures enhances their functionality and efficiency, making them a critical component of modern aerospace engineering [29].

3. Alternative fuels and propulsion systems

3.1 Electric propulsion systems

Electric propulsion (EP) systems represent a significant advancement in propulsion technology, offering benefits regarding spacecraft efficiency and sustainability. Unlike traditional chemical propulsion systems, electric propulsion uses electrical energy to accelerate propellants to high velocities, achieving greater fuel efficiency and enabling longer missions. Electric propulsion systems work by using electric power to accelerate ionized gases, creating thrust. This process results in higher exhaust velocities compared to chemical propulsion, making electric propulsion more efficient, particularly for long-duration space missions. The concept of electric propulsion has been explored and developed over the decades, leading to various types of EP systems, such as ion thrusters, Hall effect thrusters, and magnetoplasmadynamic (MPD) thrusters.

The development of electric propulsion dates back to the early twentieth century. One of the foundational works in this field is by Jahn (1964), who provided a comprehensive overview of electric propulsion principles and their potential applications. Jahn's work laid the groundwork for understanding how electric fields can be used to accelerate ions and generate thrust, a concept that has evolved significantly since its inception [30].

3.1.1 Types of electric propulsion systems

3.1.1.1 Ion thrusters

Ion thrusters use electrostatic forces to accelerate ions to high velocities. These systems are characterized by their high efficiency and specific impulse, making them ideal for deep space missions. Ion thrusters have been successfully used in missions such as NASA's Deep Space 1 and the Dawn spacecraft (**Figure 2**) [31].

3.1.1.2 Hall effect thrusters

Hall effect thrusters also use electric fields to ionize and accelerate propellants, but they rely on a different mechanism involving magnetic fields to confine the electrons. This type of thruster offers a good balance between thrust and efficiency and is widely used in Earth-orbiting satellites for station-keeping and orbital adjustments.

3.1.1.3 Magnetoplasmadynamic (MPD) thrusters

MPD thrusters use both electric and magnetic fields to accelerate plasma. These thrusters can produce higher thrust levels compared to ion and Hall effect thrusters, but they require substantial power, limiting their use to missions where high power is available [32].

Figure 2.
NASA's deep space 1 (source: NASA).

The application of electric propulsion in spacecraft design requires careful optimization to balance performance, power consumption, and mission requirements. To further enhance the efficiency and reduce the fuel mass, a holistic design process that consists of the integration of EP systems with other spacecraft subsystems for optimization of hybrid electric propulsion systems can be utilized [33]. Electric propulsion systems show extreme potential for solar system exploration, due to their capability to enable missions to distant destinations, such as the outer planets and beyond. The continuous improvement of power sources, such as solar panels and nuclear reactors, will further enhance the viability of electric propulsion for a wide range of space missions [34].

3.2 Hydrogen fuel cells

Hydrogen fuel cells emerge as a promising technology for reducing the environmental impact of aerospace operations. They offer a cleaner alternative to conventional fossil fuels, producing water as the only byproduct. Hydrogen fuel cells generate electricity through an electrochemical reaction between hydrogen and oxygen. This process involves the separation of hydrogen molecules into protons and electrons, which then travel through different paths to combine with oxygen, producing electricity, heat, and water. Hydrogen fuel cells produce zero carbon emissions during operation, significantly reducing the environmental footprint of aerospace activities [35]. Hydrogen has a high energy density, making it an efficient energy carrier for long-duration flights [36]. Fuel cell systems can be scaled to meet different power requirements, from small UAVs to large commercial aircraft [37].

3.2.1 Applications of hydrogen fuel cells

Nickel-hydrogen batteries: These batteries have been utilized in both military and commercial aerospace applications. They offer high energy density and long life cycles, making them suitable for satellite and space missions. Nickel-hydrogen batteries are known for their reliability and performance in the harsh conditions of space [38, 39].

Regenerative fuel cells: For geosynchronous space missions, hydrogen/oxygen-based regenerative fuel cells have been optimized for system mass, enhancing the efficiency of energy storage and usage in space. These systems are designed to store energy generated from solar panels during the day and use it during the night, ensuring a continuous power supply [36].

Advances in battery development: Significant progress has been made in developing advanced batteries for aerospace applications, including improvements in energy density, weight reduction, and operational efficiency. These advances are critical for the broader adoption of hydrogen fuel cells in various aerospace platforms (**Figure 3**) [37].

3.3 Biofuels

Biofuels represent a promising alternative to conventional fossil fuels, particularly in the context of reducing greenhouse gas emissions and enhancing energy security. When used in High-speed turbine locomotives, biofuels demonstrated a significant reduction in harmful emissions compared to traditional fossil fuels. Specifically, reductions in carbon dioxide (CO_2), sulfur oxides (SOx), and particulate matter (PM) were noted, contributing to better air quality and less environmental degradation.

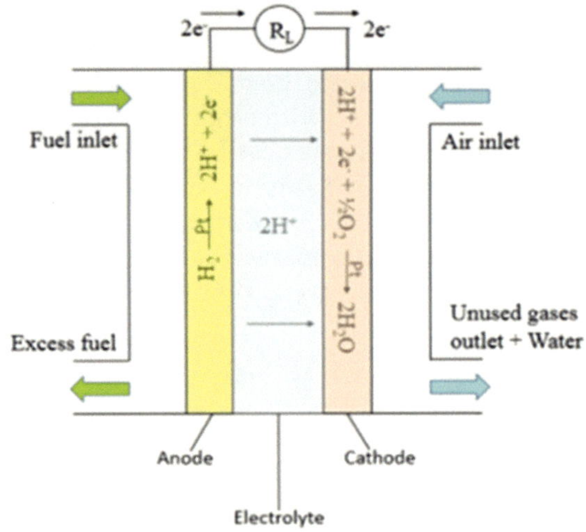

Figure 3.
Schematic illustration of a hydrogen fuel cell [40].

Biofuels show slightly lower energy content than conventional fuels, but as advances in turbine technology and fuel formulation are taking place, this issue could potentially be eliminated, making biofuels a viable alternative for high-speed applications. Also, biofuels have higher production costs due to the current scale of production and processing techniques. However, with increased adoption and technological advancements, economies of scale could lead to cost reductions over time [41].

Using a Biomass-Derived Fuel on industrial gas turbines also reveals some key performance metrics and potential benefits. The gas turbines running on biomass-derived fuels maintained comparable performance levels to those using conventional fuels, with only slight variations in efficiency and output. Turbine materials exhibited good compatibility with biomass-derived fuels, showing no significant degradation or adverse reactions, which is crucial for long-term operational sustainability. Biomass-derived fuels, being sourced from renewable organic matter, offer a sustainable alternative that can help diversify energy sources and reduce dependency on fossil fuels [42].

Bio-based fuels can be implemented in aviation through the Clean Airports Program. Biofuels could be efficiently integrated into existing aviation infrastructure with minimal modifications to engines and fuel systems, facilitating a smoother transition from conventional fuels. The use of biofuels in aviation also resulted in a marked decrease in CO_2 and other greenhouse gas emissions, aligning with global efforts to combat climate change [6].

4. Autonomous systems

4.1 Unmanned aerial vehicles

Unmanned Aerial Vehicles (UAVs), commonly known as drones, are aerial vehicles that do not require an onboard human operator. These systems have seen extensive development and application across various domains, from military operations to commercial and recreational use. UAVs consist of various structural and functional components, including airframes, propulsion systems, navigation systems, and payloads. Each component plays a critical role in the UAV's overall performance and capabilities. Advances in UAV technology, such as improved sensors, communication systems, and autonomous control, have significantly enhanced their capabilities and reliability [43]. The versatility of UAVs allows for their application in numerous fields, from military reconnaissance to civilian use. Initially, these were developed for military purposes, UAVs are now employed in agriculture, environmental monitoring, disaster management, and logistics, highlighting their adaptability to various tasks and environments. UAVs can operate in diverse environments, perform complex tasks, and gather data from inaccessible areas, making them invaluable in modern aerospace and defense strategies [44].

The design and optimization of UAVs are critical to enhancing their performance and efficiency. Factors such as aerodynamics, propulsion, weight, and stability are considered in UAV design to achieve optimal performance. Aerodynamic efficiency helps reduce fuel consumption and improve flight endurance. Advanced optimization techniques are employed to refine UAV designs, ensuring they meet mission-specific requirements and perform effectively in various conditions [45]. Effective UAV design must balance performance, reliability, and cost while addressing issues such as power consumption, communication range, and payload capacity. Understanding and improving operational constraints, such as environmental conditions and regulatory requirements, are crucial for successful UAV deployment [46].

Technological advancements continue to expand the capabilities and applications of UAV systems. Innovations in sensors, communication systems, and autonomous control have significantly enhanced UAV capabilities, allowing them to perform a wider variety of tasks with increased freedom and reliability. The widespread adoption of UAVs also brings regulatory and operational challenges that need to be addressed to ensure safe and efficient integration into airspace [47]. Cooperative control strategies enable multiple UAVs to work together, enhancing their overall effectiveness. Multiple UAVs can share information and tasks, improving efficiency and effectiveness in complex missions such as search and rescue, surveillance, and environmental monitoring. UAVs operating in coordination can cover larger areas and provide enhanced capabilities, ensuring mission success even in challenging environments [48]. UAVs are also revolutionizing 3D mapping by providing high-resolution, accurate spatial data. Equipped with advanced sensors and cameras, UAVs can capture high-resolution images and generate detailed 3D models of terrain and structures. This capability is essential for fields like archaeology, construction, and environmental monitoring. UAVs can quickly and efficiently collect spatial data from hard-to-reach areas, providing real-time insights and enhancing the accuracy of geographical analysis [49]. UAVs play a crucial role in enhancing communication networks, particularly in remote or hostile environments where traditional infrastructure is lacking or unreliable. UAVs can serve as airborne relay stations, extending the range of communication networks and ensuring robust connectivity for ground units. This capability is vital for military operations and disaster response where reliable communication is critical. The use of UAVs in communication relay enhances the security of communication channels, making it difficult for adversaries to intercept or disrupt communications. This is especially important in military applications where secure communication is paramount (**Figure 4**) [50].

4.2 Autonomous navigation and control

Autonomous navigation and control systems are essential for enabling UAVs and other autonomous vehicles to operate independently, using advanced algorithms

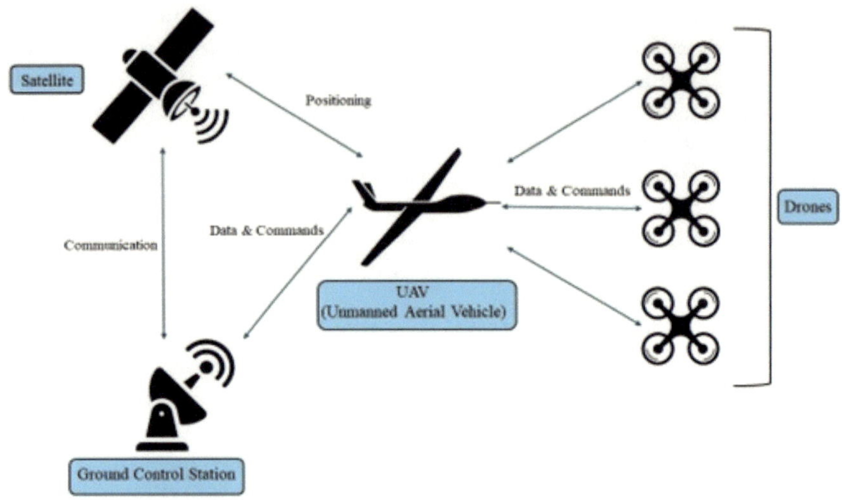

Figure 4.
Communication architecture between UAV, satellite, ground station, and drones (source: Internet).

and sensor integration. Autonomous systems integrate data from various sensors, such as GPS, LiDAR, cameras, and inertial measurement units (IMUs), to navigate and perform tasks accurately. This multi-sensor approach allows for robust real-time decision-making. By combining data from different sources, these systems can create a detailed understanding of the environment, which enhances their ability to navigate complex terrains and avoid obstacles effectively. For instance, visual relative localization techniques enable swarms of micro aerial vehicles (MAVs) to stabilize and coordinate their movements by processing visual cues from the environment [51].

Machine learning algorithms significantly enhance autonomous systems by enabling them to adapt to new environments and challenges. These algorithms allow systems to learn from records, improve their performance over time, and make informed decisions in dynamic and unpredictable environments. For example, a machine learning approach to visual perception of forest trails for mobile robots enables UAVs to navigate forest trails by learning and recognizing patterns in the visual data, which is crucial for operations in natural and unstructured environments [52].

Effective autonomous navigation requires the capability to process data in real time. This capability ensures that the system can promptly respond to environmental changes, obstacles, and other variables, maintaining operational safety and efficiency. Real-time data processing allows UAVs to make immediate adjustments to their flight paths and control parameters. Integrating low-cost sensors for real-time navigation, guidance, and control allows UAVs to process sensor data on the fly, enabling them to operate autonomously in various scenarios without delays [53].

Multi-UAV systems can work collaboratively to achieve complex tasks, such as search and rescue operations, by sharing data and coordinating their actions. This collaborative approach enhances the effectiveness and efficiency of missions by enabling multiple UAVs to cover larger areas, share insights, and assist each other. An autonomous multi-UAV system designed for search and rescue missions can utilize coordinated efforts to locate survivors, distribute resources, and relay information back to rescue teams, significantly improving mission success rates [54]. Reliable communication networks are essential for autonomous UAVs, enabling them to transmit data, receive commands, and coordinate with other UAVs. Addressing issues related to network topology, communication protocols, and data transmission reliability is necessary for seamless operation. Effective communication networks ensure that UAVs can maintain constant contact with each other and ground control stations, facilitating coordinated efforts and real-time data sharing [55].

5. Quantum computing in aerospace

Quantum computing represents a significant development in computational power and capabilities, with the potential to revolutionize various fields, including aerospace. This technology utilizes the principles of quantum mechanics to perform computations that are not possible for classical computers. The application of quantum computing in aerospace offers opportunities for advancements in optimization, simulation, and complex problem-solving.

5.1 Potential applications

5.1.1 Optimization problems

Aerospace engineering involves several optimization challenges, such as optimizing flight trajectories, designing efficient propulsion systems, and minimizing material usage in aircraft construction. Quantum computing can address these problems more effectively than classical computers due to its ability to process complex calculations simultaneously. For instance, quantum algorithms can solve optimization problems using quantum-inspired genetic algorithms, which are beneficial for complex flight control systems [56].

5.1.2 Molecular simulations

Quantum computing can significantly enhance molecular simulations, which are critical in the development of new aerospace materials and propulsion systems. For example, simulated quantum computation of molecular energies allows researchers to accurately predict the properties of new materials and their performance under various conditions. This capability can accelerate the development of advanced materials for aerospace applications [9].

5.1.3 Machine learning and data analysis

Quantum-assisted machine learning can manage large datasets and perform complex analyses more efficiently than classical approaches. This can be particularly useful for predictive maintenance and real-time analytics in aerospace operations. Quantum computing algorithms can enhance pattern recognition and anomaly detection in vast amounts of sensor data collected from aircraft and spacecraft [57].

5.2 Current research and developments

5.2.1 Variational hybrid quantum-classical algorithms

Hybrid algorithms combine classical and quantum computing methods to tackle complex problems. Variational quantum-classical algorithms, such as the Variational

Quantum Eigensolver (VQE), are currently being developed to solve problems in quantum chemistry and material science, which are directly applicable to aerospace research [10].

5.2.2 Quantum supremacy and hardware development

Researchers are working on achieving quantum supremacy, where quantum computers can perform tasks beyond the reach of classical computers. Recent advances include the development of quantum processors capable of solving specific problems more efficiently than classical systems [58]. Additionally, there is ongoing research into overcoming material challenges and improving the stability and scalability of quantum computing hardware [59].

5.2.3 Quantum chemistry and aerospace materials

The integration of quantum computing with quantum chemistry allows for accurate simulations of complex molecular structures and interactions. This capability can significantly impact the development of new aerospace materials by providing deeper insights into material properties and behaviors [60].

5.3 Future prospects

5.3.1 Enhanced computational power

As quantum computing technology matures, its ability to manage increasingly complex problems will expand. This will enable breakthroughs in aerospace engineering, including the design of more efficient propulsion systems, optimization of flight operations, and the development of advanced materials with tailored properties.

5.3.2 Integration with classical systems

The future of quantum computing in aerospace will involve hybrid systems that combine quantum and classical computing. This integration will leverage the strengths of both technologies, enhancing overall computational capabilities and allowing for more complex simulations and optimizations.

5.3.3 Broader adoption

As quantum computing hardware becomes more accessible and affordable, its adoption across the aerospace industry will increase. This will drive innovation in aerospace design, manufacturing, and operations, leading to more efficient and advanced aerospace systems.

6. Integration of Industry 4.0 and Industry 5.0

The aerospace industry is undergoing significant development through the integration of Industry 4.0 and Industry 5.0 technologies. These frameworks facilitate advanced digitalization, connectivity, and human-centric innovations, aiming to enhance efficiency, sustainability, and overall performance.

6.1 Industry 4.0 in aerospace

Industry 4.0 focuses on the digital transformation of manufacturing and related industries through smart technologies, including the Internet of Things (IoT), artificial intelligence (AI), big data analytics, and advanced robotics.

6.1.1 Digital twins and simulation

Digital twins: This technology refers to virtual copies of physical systems that enable real-time monitoring, simulation, and analysis. In aerospace, digital twins can simulate aircraft performance, predict maintenance needs, and optimize operations. For example, Rolls-Royce uses digital twins for their aircraft engines to monitor performance and predict failures, reducing downtime and maintenance costs [12].

Simulation: Advanced simulations allow for testing of various scenarios, optimizing designs, and ensuring safety. Companies like Boeing use simulations to evaluate aircraft designs under different conditions, improving safety and performance without the need for extensive physical prototypes (**Figure 5**) [61].

6.1.2 IoT and connectivity

Connected aircraft: IoT enables real-time data exchange between aircraft systems and ground operations, improving operational efficiency and passenger experience. For instance, Airbus uses IoT sensors to monitor aircraft systems in real time, providing data for predictive maintenance and operational adjustments [62].

Smart supply chains: IoT enhances the transparency and efficiency of the aerospace supply chain, tracking components through production, assembly, and delivery. This improves inventory management and reduces delays [63].

6.1.3 Advanced manufacturing techniques

Additive manufacturing: Also known as 3D printing, it allows for the creation of complex components with reduced material waste. Aerospace companies like GE Aviation use additive manufacturing to produce lighter, more efficient parts [64].

Automation and robotics: Automation improves the precision and efficiency of manufacturing processes. Robots perform repetitive and hazardous tasks, enhancing

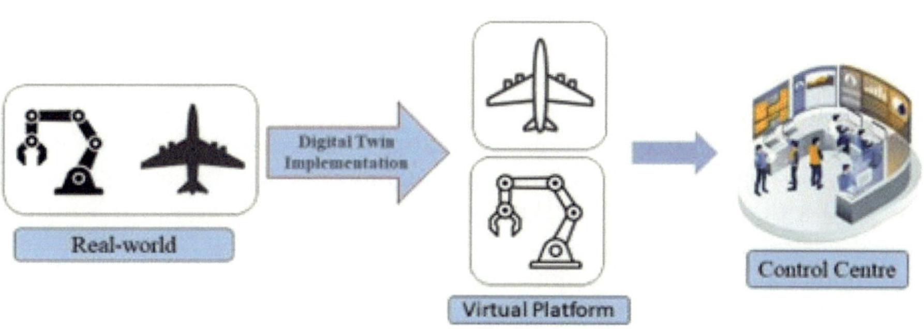

Figure 5.
Implementation of digital twin in aerospace (source: Internet).

safety and productivity. For example, robotic systems are used in aircraft assembly lines to streamline production and improve quality control [65].

6.2 Industry 5.0 in aerospace

6.2.1 Human-machine collaboration

Collaborative robots (Cobots): Cobots collaborate with human workers, enhancing productivity and safety. They are used in aerospace manufacturing to assist with complex assembly tasks, reducing human error and increasing efficiency [66].

Enhanced decision-making: AI and machine learning algorithms analyze vast amounts of data, providing insights and recommendations to human operators. This helps in decision-making processes, from design and manufacturing to maintenance and operations [67].

6.2.2 Sustainability and resilience

Sustainable practices: Industry 5.0 promotes the adoption of sustainable practices, such as using renewable materials and reducing waste. Aerospace companies are increasingly focusing on sustainability, developing eco-friendly materials and processes to minimize environmental impact [68].

Resilience and adaptability: Advanced technologies enable aerospace companies to quickly adapt to changing conditions and recover from disruptions. For example, AI-driven predictive maintenance systems can identify potential issues before they become critical, ensuring continued operation and safety [69].

6.2.3 Human-centric innovation

Enhanced worker safety and comfort: Ergonomic designs and safety systems improve the working conditions for aerospace employees. Wearable technologies monitor workers' health and safety, preventing accidents and improving overall well-being [70].

Personalized passenger experience: Industry 5.0 technologies enable a more personalized passenger experience, from customized in-flight services to seamless travel experiences. Airlines are using AI to tailor services based on passenger preferences, enhancing customer satisfaction [71].

7. Result and discussion

In this chapter, we have discussed various advanced technologies which are proven to be useful in revolutionizing the aviation industry. These advanced technologies include advanced aerospace materials, alternative fuels and propulsion systems, autonomous systems, quantum computing, and integration of Industry 4.0 and Industry 5.0 with the aviation industry.

The discussion of advanced aerospace materials reveals substantial improvements in performance and efficiency. Lightweight composite materials have demonstrated significant reductions in aircraft weight, leading to lower fuel consumption and enhanced structural integrity. High-temperature alloys are essential for enduring the extreme conditions of modern propulsion systems and hence are crucial for the development of more efficient and durable engines. Smart materials, capable of responding to environmental changes, offer opportunities for advanced adaptive systems that can enhance safety and performance. These materials can alter their properties in response to external stimuli, improving operational flexibility and resilience.

The development of electric propulsion systems is a breakthrough in reducing the aerospace industry's environmental impact. Electric propulsion can significantly decrease greenhouse gas emissions and reduce reliance on fossil fuels. Hydrogen fuel cells are emerging as a potential alternative to conventional fuels. The use of hydrogen fuel cells results in a significant reduction in carbon footprint, hence providing clean energy. Furthermore, the use of biofuels presents a promising path toward sustainable aviation. However, challenges such as fuel production, infrastructure adaptation, and cost need to be addressed to fully realize the potential of biofuels in aerospace applications.

Unmanned Aerial Vehicles (UAVs) have revolutionized aerospace operations by enabling capabilities that extend beyond human limits. The advancement in autonomous navigation and control systems allows UAVs to perform complex tasks with minimal human intervention. These systems rely on sophisticated algorithms to ensure precise and reliable operations.

Quantum computing offers transformative potential for aerospace by addressing complex problems related to optimization and simulation. Quantum computation can provide insights into molecular energies, which are valuable for developing advanced materials and propulsion systems. The development of variational hybrid quantum-classical algorithms promises to enhance computational capabilities, allowing for more accurate simulations and optimizations.

The integration of Industry 4.0 and Industry 5.0 technologies is transforming aerospace manufacturing and operations. Digital twins and simulation techniques provide real-time insights and predictive capabilities, improving design accuracy and maintenance practices. The Internet of Things (IoT) enables comprehensive monitoring and data collection, enhancing decision-making and predictive maintenance. Advanced manufacturing techniques, such as 3D printing technique streamline the manufacturing process by minimizing material wastage and reducing the time required for manufacturing which together contribute to a more efficient and sustainable manufacturing process. With the help of this technique, components with complex geometries can be manufactured with ease. These technologies collectively enhance the efficiency, flexibility, and sustainability of aerospace operations. The integration of Industry 5.0 principles, which focuses on human-centric and sustainable approaches, further enhances the flexibility and adaptability of aerospace operations.

8. Conclusion

The revolutionary advances in aerospace technology discussed in this chapter include quantum computing, autonomous systems, alternative propulsion systems and fuels, advanced materials, and the integration of Industry 4.0 and Industry 5.0. Lightweight composites reduce aircraft weight, improving fuel efficiency and structural integrity, while high-temperature alloys are crucial for modern engines. Smart materials offer adaptive properties for enhanced safety and performance. Electric propulsion reduces greenhouse gas emissions and reliance on fossil fuels, and hydrogen fuel cells offer a clean energy source. Biofuels provide a sustainable aviation alternative, though challenges in production and infrastructure remain. UAVs extend operational capabilities beyond human limits, with advanced autonomous navigation and control systems ensuring precise and reliable operations with minimal human intervention. Quantum computing enhances aerospace through advanced optimization and simulation, offering insights into molecular energies and improving computational capabilities with variational hybrid algorithms. Industry 4.0 technologies like digital twins and IoT improve design accuracy, maintenance, and decision-making. Industry 5.0 focuses on human-centric and sustainable approaches, enhancing flexibility, efficiency, and sustainability in aerospace manufacturing and operations. The results show that these developments provide notable gains in sustainability, efficiency, and performance for a variety of aerospace applications.

Acknowledgements

I must thank my wife Mrs. Rashmi Srivastava and my son Arjun Srivastava for supporting and encouraging me always. The emotional support required to do research work is often underestimated. I would like to thank God Almighty for giving me the opportunity and guidance to achieve my goal and succeed in this part.

Conflict of interest

The authors declared that they have no conflict of interests in this work.

Author details

Abhay Dhasmana and Sachin Srivastava*
Department of Aerospace Engineering, Uttaranchal Institute of Technology, Uttaranchal University, Dehradun, Uttarakhand, India

*Address all correspondence to: sachinrash2004@gmail.com

References

[1] Farag MM. Materials and Process Selection for Engineering Design. CRC Press; 2020. DOI: 10.1201/9781003006091

[2] Mangalgiri PD. Composite materials for aerospace applications. Bulletin of Materials Science. 1999;**22**:657-664. DOI: 10.1007/BF02749982

[3] Gandhi MV, Thompson BD. Smart Materials and Structures. Springer Science & Business Media; 1992

[4] Fornaro E, Cardone M, Dannier A. A comparative assessment of hybrid parallel, series, and full-electric propulsion systems for aircraft application. IEEE Access. 2022;**10**:28808-28820. DOI: 10.1109/ACCESS.2022.3158372

[5] Revankar ST, Majumdar P. Fuel Cells: Principles, Design, and Analysis. CRC Press; 2014

[6] Shauck ME, Zanin MG. Implementation of Alternative Bio-Based Fuels in Aviation: The Clean Airports Program (No. CONF-970856-3). Waco, TX (United States): Baylor Univ., Renewable Aviation Fuels Development Center; 1997

[7] Lozano R, editor. Unmanned Aerial Vehicles: Embedded Control. John Wiley & Sons; 2013

[8] Nerode A, Kohn W. An autonomous systems control theory: An overview. In: IEEE Symposium on Computer-Aided Control System Design. IEEE; 1992. pp. 204-210

[9] Aspuru-Guzik A, Dutoi AD, Love PJ, Head-Gordon M. Simulated quantum computation of molecular energies. Science. 2005;**309**(5741):1704-1707. DOI: 10.1126/science.1113479

[10] McClean JR, Romero J, Babbush R, Aspuru-Guzik A. The theory of variational hybrid quantum-classical algorithms. New Journal of Physics. 2016;**18**(2):023023. DOI: 10.1088/1367-2630/18/2/023023

[11] Preskill J. Quantum computing in the NISQ era and beyond. Quantum. 2018;**2**:79. DOI: 10.22331/q-2018-08-06-79

[12] Tao F, Zhang H, Liu A, Nee AY. Digital twin in industry: State-of-the-art. IEEE Transactions on Industrial Informatics. 2018;**15**(4):2405-2415. DOI: 10.1109/TII.2018.2873186

[13] Stanton I, Munir K, Ikram A, El-Bakry M. Predictive maintenance analytics and implementation for aircraft: Challenges and opportunities. Systems Engineering. 2023;**26**(2):216-237. DOI: 10.1002/sys.21651

[14] Berman B. 3D printing: The new industrial revolution. IEEE Engineering Management Review. 2013;**41**(4):72-80. DOI: 10.1109/EMR.2013.6693869

[15] Elangovan U. Industry 5.0: The Future of the Industrial Economy. CRC Press; 2021. DOI: 10.1201/9781003190677

[16] Steinberg MA. Materials for aerospace. Scientific American. 1986;**255**(4):66-73

[17] Bunk WJG. Aerospace materials, situation and perspectives. In: Advanced Aerospace Materials. Berlin, Heidelberg: Springer Berlin Heidelberg; 1992. pp. 1-20. DOI: 10.1007/978-3-642-50159-3_1

[18] Dufton PW. Lightweight Thermoset Composites: Materials in Use, Their Processing and Applications. iSmithers Rapra Publishing; 2000

[19] Salkind M. Aerospace materials research opportunities. Advanced Materials. 1989;**1**(5):157-164. DOI: 10.1002/adma.19890010506

[20] Perrut M, Caron P, Thomas M, Couret A. High temperature materials for aerospace applications: Ni-based superalloys and γ-TiAl alloys. Comptes Rendus Physique. 2018;**19**(8):657-671. DOI: 10.1016/j.crhy.2018.10.002

[21] Antolovich SD, Busso EP, Skelton P, Telesman J. High temperature materials for aerospace applications. Materials at High Temperatures. 2016;**33**(4-5): 289-290. DOI: 10.1080/09603409.2016.1206294

[22] Das SK, Davis LA. High performance aerospace alloys via rapid solidification processing. Materials Science and Engineering. 1988;**98**:1-12. DOI: 10.1016/0025-5416(88)90116-4

[23] Gogia AK. High-temperature titanium alloys. Defence Science Journal. 2005;**55**(2):149-173. DOI: DOI

[24] Bochenek K, Basista M. Advances in processing of NiAl intermetallic alloys and composites for high temperature aerospace applications. Progress in Aerospace Sciences. 2015;**79**:136-146. DOI: 10.1016/j.paerosci.2015.09.003

[25] Bandyopadhyay K. Smart materials and aerospace structures. In: Indo-Russian Workshop on Micromechanical Systems. Vol. 3903. SPIE; 1999. pp. 312-324. DOI: 10.1117/12.369472

[26] Basheer AA. Advances in the smart materials applications in the aerospace industries. Aircraft Engineering and Aerospace Technology. 2020;**92**(7):1027-1035. DOI: 10.1108/AEAT-02-2020-0040

[27] Wang W, Xiang Y, Yu J, Yang L. Development and prospect of smart materials and structures for aerospace sensing systems and applications. Sensors. 2023;**23**(3):1545. DOI: 10.3390/s23031545

[28] Boller C. Smart structures and materials in aerospace applications of next generation. WEIGHT 2650 KG (MAX 4170), 133; 1997

[29] Nathal MV, Stefko GL. Smart materials and active structures. Journal of Aerospace Engineering. 2013;**26**(2):491-499. DOI: 10.1061/(ASCE)AS.1943-5525.000031

[30] Jahn RG. Electric propulsion. American Scientist. 1964;**52**(2):207-217

[31] Martinez-Sanchez M, Pollard JE. Spacecraft electric propulsion-an overview. Journal of Propulsion and Power. 1998;**14**(5):688-699. DOI: 10.2514/2.5331

[32] Wilbur PJ, Jahn RG, Curran FC. Space electric propulsion plasmas. IEEE Transactions on Plasma Science. 1991;**19**(6):1167-1179. DOI: 10.1109/27.12503910.1109/27.125039

[33] Assanis D, Delagrammatikas G, Fellini R, Filipi Z, Liedtke J, Michelena N, et al. Optimization approach to hybrid electric propulsion system design. Journal of Structural Mechanics. 1999;**27**(4):393-421. DOI: 10.1080/08905459908915705

[34] Brophy JR, Noca M. Electric propulsion for solar system exploration. Journal of Propulsion and Power. 1998;**14**(5):700-707. DOI: 10.2514/2.5332

[35] Appleby AJ. Fuel cells and hydrogen fuel. International Journal of

Hydrogen Energy. 1994;**19**(2):175-180. DOI: 10.1016/0360-3199(94)90124-4

[36] Hauff S, Bolwin K. System mass optimization of hydrogen/oxygen based regenerative fuel cells for geosynchronous space missions. Journal of Power Sources. 1992;**38**(3):303-315. DOI: 10.1016/0378-7753(92)80120-Z

[37] Fox CL, Marble SC. Advances in the Development of Batteries for Aerospace. 1997

[38] Coates DK, Fox CL, Miller LE. Hydrogen-based rechargeable battery systems: Military, aerospace, and terrestrial applications—I. Nickel-hydrogen batteries. International Journal of Hydrogen Energy. 1994;**19**(9):743-750. DOI: 10.1016/0360-3199(94)90238-0

[39] Caldwell DB, Coates DK, Fox CL, Miller LE. Commercial aerospace and terrestrial applications of nickel-hydrogen batteries. In: AIP Conference Proceedings. Vol. 361. American Institute of Physics; 1996. pp. 199-204

[40] Slaughter G, Kulkarni T. Enzymatic glucose biofuel cell and its application. Journal of Biochips & Tissue Chips. 2015;**5**(1):1. DOI: 10.4172/2153-0777.1000110

[41] Pier JR. Comparisons of biofuels in high-speed turbine locomotives: Emissions, energy use, and cost. Transportation Research Record. 1999;**1691**(1):24-32. DOI: 10.3141/1691-04

[42] Andrews RG, Fuleki D, Zukowski S, Patnaik PC. Results of Industrial Gas Turbine Tests Using a Biomass Derived Fuel. 1997

[43] Fahlstrom PG, Gleason TJ, Sadraey MH. Introduction to UAV Systems. John Wiley & Sons; 2022

[44] Van Blyenburgh P. UAVs: An overview. Air and Space Europe. 1999;**1**(5):43-47. DOI: 10.1016/S1290-0958(00)88869-3

[45] Kontogiannis SG, Ekaterinaris JA. Design, performance evaluation and optimization of a UAV. Aerospace Science and Technology. 2013;**29**(1):339-350. DOI: 10.1016/j.ast.2013.04.005

[46] Torun E. UAV Requirements and design consideration. In: Advances in Vehicle Systems Concepts and Integration. 2000. p. 108

[47] Fan B, Li Y, Zhang R, Fu Q. Review on the technological development and application of UAV systems. Chinese Journal of Electronics. 2020;**29**(2):199-207. DOI: 10.1049/cje.2019.12.006

[48] Ryan A, Zennaro M, Howell A, Sengupta R, Hedrick JK. An overview of emerging results in cooperative UAV control. In: 2004 43rd IEEE Conference on Decision and Control (CDC) (IEEE Cat. No. 04CH37601). Vol. 1. IEEE; 2004. pp. 602-607

[49] Nex F, Remondino F. UAV for 3D mapping applications: A review. Applied Geomatics. 2014;**6**:1-15. DOI: 10.1007/s12518-013-0120-x

[50] Pinkney MF, Hampel D, DiPierro S. Unmanned aerial vehicle (UAV) communications relay. In: Proceedings of MILCOM'96 IEEE Military Communications Conference. Vol. 1. IEEE; 1996. pp. 47-51

[51] Saska M, Vakula J, Přeućil L. Swarms of micro aerial vehicles stabilized under a visual relative localization. In: 2014 IEEE International Conference on Robotics and Automation (ICRA). IEEE; 2014. pp. 3570-3575

[52] Giusti A, Guzzi J, Cireşan DC, He FL, Rodríguez JP, Fontana F, et al.

A machine learning approach to visual perception of forest trails for mobile robots. IEEE Robotics and Automation Letters. 2015;**1**(2):661-667. DOI: 10.1109/LRA.2015.2509024

[53] Kim JH, Sukkarieh S, Wishart S. Real-time navigation, guidance, and control of a UAV using low-cost sensors. In: Field and Service Robotics: Recent Advances in Research and Applications. Berlin, Heidelberg: Springer Berlin Heidelberg; 2006. pp. 299-309. DOI: 10.1007/10991459_29

[54] Scherer J, Yahyanejad S, Hayat S, Yanmaz E, Andre T, Khan A, et al. An autonomous multi-UAV system for search and rescue. In: Proceedings of the First Workshop on Micro Aerial Vehicle Networks, Systems, and Applications for Civilian Use. 2015. pp. 33-38. DOI: 10.1145/2750675.2750683

[55] Gupta L, Jain R, Vaszkun G. Survey of important issues in UAV communication networks. IEEE Communications Surveys & Tutorials. 2015;**18**(2):1123-1152. DOI: 10.1109/COMST.2015.2495297

[56] Qi BIAN, Nener B, Xinmin WANG. A quantum inspired genetic algorithm for multimodal optimization of wind disturbance alleviation flight control system. Chinese Journal of Aeronautics. 2019;**32**(11):2480-2488. DOI: 10.1016/j.cja.2019.04.013

[57] Perdomo-Ortiz A, Benedetti M, Realpe-Gómez J, Biswas R. Opportunities and challenges for quantum-assisted machine learning in near-term quantum computers. Quantum Science and Technology. 2018;**3**(3):030502. DOI: 10.1088/2058-9565/aab859

[58] Boixo S, Isakov SV, Smelyanskiy VN, Babbush R, Ding N, Jiang Z, et al. Characterizing quantum supremacy in near-term devices. Nature Physics. 2018;**14**(6):595-600. DOI: s41567-018-0124-x

[59] De Leon NP, Itoh KM, Kim D, Mehta KK, Northup TE, Paik H, et al. Materials challenges and opportunities for quantum computing hardware. Science. 2021;**372**(6539):eabb2823. DOI: 10.1126/science.abb2823

[60] Cao Y, Romero J, Olson JP, Degroote M, Johnson PD, Kieferová M, et al. Quantum chemistry in the age of quantum computing. Chemical Reviews. 2019;**119**(19):10856-10915. DOI: 10.1021/acs.chemrev.8b00803

[61] Glaessgen E, Stargel D. The digital twin paradigm for future NASA and US air force vehicles. In: 53rd AIAA/ASME/ASCE/AHS/ASC Structures, Structural Dynamics and Materials Conference 20th AIAA/ASME/AHS Adaptive Structures Conference 14th AIAA. 2012. p. 1818

[62] Gubbi J, Buyya R, Marusic S, Palaniswami M. Internet of things (IoT): A vision, architectural elements, and future directions. Future Generation Computer Systems. 2013;**29**(7):1645-1660. DOI: 10.1016/j.future.2013.01.010

[63] Atzori L, Iera A, Morabito G. The internet of things: A survey. Computer Networks. 2010;**54**(15):2787-2805. DOI: 10.1016/j.comnet.2010.05.010

[64] Gibson I, Rosen DW, Stucker B, Khorasani M, Rosen D, Stucker B, et al. Additive Manufacturing Technologies. Vol. 17. Cham, Switzerland: Springer; 2021. pp. 160-186. DOI: 10.1007/978-3-030-56127-7

[65] Bogue R. The growing use of robots by the aerospace industry. Industrial Robot: An International Journal. 2018;**45**(6):705-709. DOI: 10.1108/IR-08-2018-0160

[66] Cherubini A, Passama R, Crosnier A, Lasnier A, Fraisse P. Collaborative manufacturing with physical human–robot interaction. Robotics and Computer-Integrated Manufacturing. 2016;**40**:1-13. DOI: 10.1016/j.rcim.2015.12.007

[67] Cheng GJ, Liu LT, Qiang XJ, Liu Y. Industry 4.0 development and application of intelligent manufacturing. In: 2016 International Conference on Information System and Artificial Intelligence (ISAI). IEEE; 2016. pp. 407-410

[68] Jasiński D, Meredith J, Kirwan K. A comprehensive framework for automotive sustainability assessment. Journal of Cleaner Production. 2016;**135**:1034-1044. DOI: 10.1016/j.jclepro.2016.07.027

[69] Lee J, Bagheri B, Kao HA. A cyber-physical systems architecture for industry 4.0-based manufacturing systems. Manufacturing Letters. 2015;**3**:18-23. DOI: 10.1016/j.mfglet.2014.12.001

[70] Lim HJ, Black TR, Shah SM, Sarker S, Metcalfe J. Evaluating repeated patient handling injuries following the implementation of a multi-factor ergonomic intervention program among health care workers. Journal of Safety Research. 2011;**42**(3):185-191. DOI: 10.1016/j.jsr.2011.05.002

[71] Tussyadiah I. A review of research into automation in tourism: Launching the annals of tourism research curated collection on artificial intelligence and robotics in tourism. Annals of Tourism Research. 2020;**81**:102883. DOI: 10.1016/j.annals.2020.102883

Section 2

New Technologies in Aerospace Engineering

Chapter 4

Beyond the Blue Skies: A Comprehensive Guide for Risk Assessment in Aviation

Leila Halawi, Mark Miller and Sam Holley

Abstract

Risk assessment in aviation is a critical process that safeguards the safety and reliability of operations. Aviation operations encompass inherent risks, from mechanical failures to human errors and environmental factors. The significance of these risks may be severe, leading to accidents, injuries, and loss of life. Recognizing and mitigating risks is supreme in this dynamic environment, where emerging technologies and innovation constantly reshape this industry. This chapter includes an in-depth explanation of risk management and analysis, leading to the core elements of risk assessment specifically for aviation operations. We will describe the process and explore some of the applications and methodologies for Risk Assessment. Lastly, we will discuss safety management systems followed by proactive risk analysis using explainable artificial intelligence (XAI), which can enhance aviation safety and inform engineering design decisions.

Keywords: risk assessment, risk mitigation and management, safety management systems [SMS], explainable artificial intelligence [XAI], aviation operations, human factors

1. Introduction

Aviation operations combine different specialties and actions to guarantee air transport systems' safety, efficiency, and reliability. Aviation operations proactively control and decrease hazards by methodically applying risk assessment, thus improving largely safety and efficiency. In this context, operations may include government regulators, company operation centers, equipment consortia, training for operators, and related parties with vested interests in the safety and efficiency of associated operations.

Modern risk management theory embraces a holistic approach, recognizing the importance of addressing all risks, regardless of their familiarity or ease of quantification.

Risk assessment, sometimes described as risk management analysis (FFA RM Hbk), typically is regarded as the second step in the risk management process and is the systematic process of recognizing, evaluating, and mitigating risks to minimize

Figure 1.
Risk assessment steps.

their impact on operations, which is essential in aviation to protect passengers, crew, and aircraft. It applies to flight operations, maintenance, safety management systems, regulatory compliance, and emergency preparedness. It encompasses five steps highlighted in **Figure 1**.

Operational risk assessment encompasses the comprehensive process of identifying, analyzing, and evaluating risks during the operational phase.

This chapter provides a solid foundation for risk assessment concepts, components, current and emerging approaches, and processes. This chapter focuses only on the operational risks.

The remainder of this chapter follows a structured organization. Section 2 delves into understanding and explaining fundamental concepts and terminology. Section 3 explores risk and hazard identification. Section 4 delves into risk assessment. Section 5 focuses on risk mitigation, monitoring, and review. Section 6 covers Safety Management Systems (SMS), Section 7 explores the integration of Human Factors into SMS, and Section 8 examines proactive risk analysis using Explainable Artificial Intelligence (XAI). Section 9 serves as the culmination of the chapter, summarizing key insights and findings.

2. Fundamental concepts and terminology

In the 1930s, the aviation industry led in technological advancements but relied on trial-and-error safety protocols, known as "fly–fix–fly" [1]. Overall, from the 1930s to 1999, significant advancements in aviation safety were driven by technology, an increasing safety focus, and the establishment of regulations and risk assessment practices. As air travel expanded, efforts to prevent accidents intensified. Investigations into major incidents spurred improvements in aircraft design, pilot training, and maintenance. Regulatory bodies like the Civil Aeronautics Administration (CAA) and later the Federal Aviation Administration (FAA) were established to set comprehensive safety standards. A significant application of risk management in aviation operations was implemented by the US Army Rotary Wing with Operational Risk Management in 1990, which was later adopted by the Department of Defense in 2000 and eventually the FAA. In 2001, a shift to proactive safety occurred, with ICAO initiating the Safety Management Systems approach recommendation for use in international aviation for specific aviation sectors. In 2006, ICAO required most commercial aviation service providers to implement SMS, with the FAA adopting SMS for airports in 2010 and commercial airlines in 2018. The FAA has become proactive with risk assessment through the growth of operational risk management and SMS risk assessment methodologies to identify and address potential hazards before they materialize, emphasizing preventive measures.

Key organizations in aviation risk assessment include the International Civil Aviation Organization (ICAO) and the International Air Transport Association (IATA). the Flight Safety Foundation (FSF), Civil Aviation Authorities (CAAs), the European Aviation Safety Agency [EASA], the US Federal Aviation Administration [FAA]. the UK Civil Aviation Authority [CAA], and the European Organization for the Safety of Air Navigation [EUROCONTROL].

ICAO doc 9859 [2] is the ICAO's guidance material for SMS implementation. The safety management manual emphasizes safety principles and data-driven decision-making. It focuses on safety intelligence, hazard identification through past incidents, and safety performance management. It discusses accident causation, safety culture, and safety performance management. Safety risk index ranges determine recommended actions for risk mitigation. Similar to the FAA approach (see below), a fifth category of likelihood is added - extremely improbable. The categories for severity differ from the FAA version and are named catastrophic, hazardous, major, minor, and negligible.

IATA published the Integrated Risk and Resilience Management Manual (IRRM), offering comprehensive guidance for aviation entities to integrate risk management and enhance emergency capabilities [3].

The FAA Risk Management Handbook emphasizes risk management and pilot safety in the aviation environment; it offers guidance for pilots to manage risk and describes the severity levels of outcomes; it focuses on risk management, workload, and safety in aviation operations. The Handbook describes assessment as risk management analysis. A flight risk assessment tool (FRAT) results in a score that covers three ranges: low (green), medium (yellow), and high (red). FRAT is a much simpler version of an Operational Readiness Evaluation. Risk assessment is characterized as the composite of likelihood (probability) and severity (consequences) of an outcome expressed as RA = P + S. The levels of likelihood include probable, occasional, remote, and improbable. For severity, the levels are catastrophic, critical, marginal, and negligible [4].

The EUROCONTROL safety regulatory requirement [ESARR4] identifies risks in air traffic management (ATM) for systematic management. It focuses on risk assessment and mitigation in ATMs. It addresses software safety objectives for risk mitigation. It enhances safety in ATMs through risk assessment and mitigation and ensures systematic hazard identification and management within approved safety levels. The Safety Regulation Commission prepares it for ATM system changes [5].

The Airport Cooperative Research Program (ARCP) Report 50, sponsored by the FAA, included guidance on improved models for risk assessment of runway safety areas. The report focuses on risk assessment models for runway safety areas and highlights variables impacting the consequences of overruns and accidents. It also provides rational, risk-based assessment tools for airport safety improvements. The report aims to solve aviation operating problems and introduce innovations. The report reviews operational experience to develop a functional hazard analysis. It also assesses runway safety risks based on obstacle proximity and speed [6].

The FAA's comprehensive website on SMS covers ICAO policy, US policy, regulatory requirements, commercial aviation, air traffic control, and airports. The FAA Office of Airports Safety Management Systems, desk reference version 2.1. 2023 focuses on risk assessment in airport construction projects and highlights safety measures and project coordination for aviation facilities [7]. It is an adaptation of the operational safety assessment [OSA] from the FAA System Safety Handbook with mitigation measures, a two-step process starting with identifying an airport system's ground and air elements and an operational hazard assessment. OSA and comparative safety assessment [CSA] are the main methods discussed in the reference. CSA is a risk assessment tool that defines the severity and likelihood of risks associated with each alternative under consideration. It also describes airport systems and facilities for safety risk management assessments.

2.1 Definition and terminology

This section has delineated several frequently used terms in risk assessment. The objective is to establish a standardized terminology that facilitates describing various components involved in a risk analysis.

Risk Definition: The traditional definition of risk is the possibility of loss or injury. The Risk and Insurance Management Society defines risk as an uncertain future outcome that can improve or worsen your position.

According to the committee of sponsoring organizations of the Treadway Commission (COSO), the risk is the possibility that an event will occur and adversely affect the achievement of objectives. According to ISO 31000:2015 from the International Organization for Standardization, risk is defined as "the effect of uncertainty on objectives [8]." This definition reflects contemporary perspectives on risk, acknowledging that it can lead to positive and negative outcomes. Risk is the combined answers to three critical questions: (1) What can go wrong? (2) How likely is it to happen? Moreover, and (3) What are the potential consequences?

2.2 Related terms

Other common relevant terms are listed below:

- Accident: an abrupt, undesired, and unanticipated event or sequence that harms people, the environment, or other tangible assets.

- Barriers: safeguards, protective layers, defenses, controls, or countermeasures are measures implemented to mitigate risks and prevent unwanted events from occurring or escalating.

- Consequence: the significance of an event's impact.

- Event: event" denotes a forthcoming happening.

- Hazard: A resource or precondition that can instigate damage, alone or in combination with other elements.

DOI: http://dx.doi.org/10.5772/intechopen.1006687

- Incident: an unexpected undesirable and unintentional occurrence or series of events that can realistically instigate damage to one or more resources but eventually did not cause sizable damage.

- ISO 31000: a global standard established by the International Organization for Standardization (ISO) for risk management, offering instructions and rules for creating, applying, and sustaining efficient risk management processes within companies [8].

- Likelihood: a synonym for the probability of an event, a number between 0 and 1 (or between 0 and 100%) that indicates how likely the event is to occur in a specific situation. This probability is written as PPP. If P = 1P = 1P = 1, the event is certain to happen, while if P = 0P = 0P = 0, the event is certain not to occur.

- Operational Risk: "the risk of loss stemming from lacking or failed internal processes, people, systems, or external events." ISO 31000 defines it as "any event that affects an organization's objectives."Operational risk is a pure risk that arises from people, processes, systems, and controls. This includes a broad spectrum of risks attributable to human factors, IT risks, management oversight, and business processes [8].

- Potential Cause: some element, circumstance, or incident that can generate or influence a risk occasion.

- Potential Outcome: the probable effects or impacts of a risk event. This comprises both negative and positive outcomes.

- Prevention: actions, tactics, and schemes to moderate and lessen the probability of a risk case.

- Recovery: activities and procedures started to reestablish usual tasks and diminish the bearing following a risk event.

- Safety Management Systems (SMS): a formal, structured process that helps organizations manage safety risk in the workplace.

- Threat: A broad classification encompassing actions or events that can inflict harm on an asset.

3. Risk and hazards identification

Hazard risks are distinguished by a key characteristic: pure risks with only negative potential outcomes, such as fire, theft, or catastrophe. Hazards risks arise from property, liability, or personnel loss exposure. Identifying these risks within an organization is crucial for adequately preparing and mitigating any possible negative consequences. Operational risks are also pure risks from people, processes, systems, and controls.

Risk identification entails identifying the assets, threats, controls, vulnerabilities, and consequences. This step aims to answer what can go wrong. It aims to identify all relevant hazards and hazardous events during intended use, foreseeable misuse, and interactions with the study object while also describing their characteristics, presence within the object, and potential initiating events. Once identified, the probability and severity of the events must be assessed.

Hazard identification and risk assessment methods are crucial in aviation. Once assets, threats, and vulnerabilities are identified, organizations assess the risks quantitatively or qualitatively, prioritizing those with the greatest potential impact. Several approaches for conducting a risk assessment are available.

Some methods include checklists and brainstorming, which involves lists of generic hazards or hazardous events to determine their potential occurrence, location, and impact on the study object and provide initial insights.

Preliminary Hazard Analysis (PHA) is a straightforward method in design phases, identifying hazards with results updated as more detailed risk analyses progress; it is often sufficient for basic systems and referred to as Hazard Identification (HAZID).

Job Safety Analysis (JSA), similar to PHA, enhances safety awareness among personnel by reviewing work procedures just before tasks, ensuring readiness, and mitigating risks effectively.

Failure Modes, Effects, and Criticality Analysis (FMECA) focuses on identifying potential failure modes of system components, their causes, and the resulting impacts on the overall system reliability.

Hazard and Operability (HAZOP) studies, developed for process plants, employ structured brainstorming with guidewords to detect deviations and hazardous scenarios and are widely adopted in initial design phases and subsequent system modifications.

Structured What-If Technique (SWIFT) involves experts posing and answering structured "what-if" questions in a brainstorming format, serving as a simplified approach akin to HAZOP for various system types.

Additional hazard identification techniques are prevalent in aviation, such as predictive safety analysis using machine learning techniques and algorithms to analyze vast amounts of data from various sources, including flight recorders, weather systems, maintenance logs, and even pilot reports. It looks for subtle patterns or anomalies that might indicate potential safety issues. It is a data-driven proactive approach that can identify issues before they become critical.

Flight data monitoring analyzes aircraft systems data and continuously records numerous parameters during flight operations. This data is then analyzed to identify trends, exceedances, or deviations from standard procedures. It offers objective data on flight operations, helps identify systemic issues or trends, and can be used for targeted training and procedure improvements.

Safety management systems (SMS) are a comprehensive approach that includes risk assessment, hazard reporting, and safety assurance. SMS emphasizes continuous improvement and a proactive approach to safety rather than just reacting to incidents after they occur. SMS is an organized approach to managing safety, including systematic procedures, practices, and policies. It typically includes four components: safety policy, risk management, safety assurance, and safety promotion. It encourages a positive safety culture and facilitates continuous improvement in safety performance. However, implementing it fully can be resource-intensive, requiring commitment at all organizational levels, and may need more support due to perceived bureaucracy.

Human Factors Analysis and Classification System (HFACS) is the framework for investigating and analyzing human factors in aviation accidents. HFACS is a comprehensive framework that categorizes human error at four levels: organizational influences, unsafe supervision, preconditions for unsafe acts, and unsafe acts. It helps investigators and safety analysts understand the full context of human errors in aviation incidents. It provides a structured approach to analyzing human factors, Helps identify systemic issues that contribute to human error, and Facilitates the development of targeted interventions. However, it requires thorough training to be used effectively, can be time-consuming to apply in complex incidents, and may oversimplify some aspects of human behavior.

Advanced simulation and virtual reality training may also identify potential hazards in various scenarios without real-world risk.

Bow-tie analysis is a visual risk assessment tool that helps identify potential causes and consequences of hazards. It is valuable for conceptualizing and analyzing risks. A recognized hazardous event is positioned at the center of the diagram, with its causes depicted on the left side and the consequences on the right side. It is widely applied to assess the risks of aviation combining—two analyses (fault tree analysis and event tree analysis). The fault tree portion of the system deals with possible causes of the system's top event, and the event tree portion is related to post-event analysis. It helps identify where additional safety measures might be needed. It also helps communicate complex risk scenarios to various stakeholders.

4. Risk analysis, evaluation: Risk assessment

Risk Assessment is a process that involves planning, preparing, executing, and reporting a risk analysis, followed by evaluating the results against established risk acceptance criteria. As illustrated in **Figure 2**, the process begins with conducting a risk analysis to identify and describe relevant accident and incident scenarios and their likelihoods, collectively defining the risk. The second phase, risk evaluation, involves comparing the risk the analysis determines with the risk acceptance criteria.

Risk assessment, according to ISO 31000 [8], encompasses three essential steps:

1. *Risk Identification*: This initial phase involves pinpointing potential organizational risks.

2. *Risk Analysis*: This step assesses the likelihood of each identified risk occurring and its potential impact.

3. *Risk Evaluation*: The risk analysis results are compared against the organization's criteria to determine if further action is required.

Figure 2.
Risk assessment.

Risk analysis is a systematic study that identifies and describes potential issues, their causes, probabilities, and possible consequences. While various approaches proliferate, there are four common risk assessment techniques: a risk matrix, a decision tree, failure modes and effects analysis, and the bowtie method. The assessment should be undertaken only by competent staff who understand the process, the hazard and risk, and the activity that forms the risk.

Before initiating the risk analysis, it is essential to identify the relevant assets to consider. These assets, such as people, the environment, and reputation, are outlined within the study's scope. Risk analysis, a crucial part of a risk assessment, involves understanding the nature of the risks and determining their levels.

Risk assessment includes risk analysis and risk evaluation. A broad array of information is essential as input for any risk assessment, especially for the data requirements of quantitative risk assessments from descriptive data (technical, operational, production, maintenance, environmental, meteorological, safety, and shareholder) to probabilistic data (accident data, hazard data, natural event, human error, reliability data, and consequence data). Under directive 376/2014 in the EU, all civil aviation incidents and accident data must be collected, reported, and analyzed, supported by ECCAIRS 2 to enhance safety. The FAA manages this in the USA through the ASIAS program for sharing and analyzing safety data, while the NTSB investigates incidents to improve aviation safety. Also, The International Civil Aviation Organization (ICAO) maintains a database and publishes accident statistics on its website at [http://www.icao.int.]. Reliability and availability data in aviation is collected and analyzed through industry-specific solutions, regulation-driven reporting, and internal programs.

Risk evaluations follow where you evaluate the risk. In many cases, we present the results of a risk assessment in a traditional report and other formats, including brochures, simplified versions of the report, or presentations. The chosen format should be tailored to the stakeholders and the significance of the decisions the risk assessment is intended to support. A potential structure for a risk assessment report includes a title page, a disclaimer, an executive summary [with introduction, objectives, and limitations], an analysis approach, main conclusions and recommendations, document references, acronyms and glossary, study team, appendices and wrapping up with a section for comments.

5. Risk mitigation, monitoring and review

Risk Mitigation, a subset of risk treatment, refers to actions taken to reduce the likelihood or impact of a risk. It is one of the strategies that can be employed within the broader scope of risk treatment. Risk treatment alternatives entail risk reduction, modification, retention, avoidance, sharing, and acceptance, which means accepting the risk. Risk avoidance involves eliminating risk by changing business practices or operations.

Risk monitoring and review are crucial elements of the risk management process as defined by ISO 31000 [8]. These components ensure that the organization's risk management practices remain effective and adaptable to changes in the risk environment. Risk monitoring is an ongoing process that tracks the status of identified risks and the progress of risk treatment options. Risk review entails a periodic and systematic evaluation of the risk management process to assess its suitability, adequacy, and effectiveness.

Monitoring and Reviewing Risk involves ensuring the effectiveness and continued functionality of risk reduction measures, such as verifying procedures, equipment placement, and personnel training. It includes regular inspections and retraining to maintain technical systems and emergency response capabilities. Additionally, it monitors safety performance and risk levels through performance indicators to detect and address negative trends early.

6. Safety management systems [SMS]

SMS has become the standard for applying risk management in the aviation industry over the last 20 years. This is because it is scientifically accurate through a standard risk assessment chart (using mathematical probability), can be flexibly tailored for any type and size of aviation-related organization, and, most importantly, allows for data-driven proactive design. While consisting of the four SMS pillars of Risk Management, Assurance, Promotion, and Policy, SMS is truly centered around the Risk Management Pillar. Assurance, Promotion, and Policy are key areas to support Risk Management. The Risk Management process in any SMS needs to include hazard identification, risk assessment, and mitigation.

In contrast, the Risk Management process is monitored and updated with the Assurance pillar by evaluating the effectiveness of risk mitigation control strategies and helping identify new hazards. The policy shows management's support for SMS while defining the methods, processes, and organizational structure required to utilize SMS. The Promotion is about establishing an SMS culture through leadership and the training and communication required to implement and maintain the SMS running well.

SMS utilizes risk management as its cornerstone as a new millennium safety program. This is important to any organization because it places safety and risk management into a scientific realm using accurate risk assessment to convert the hazard into a mathematical probability standardized for the whole organization. This avoids pitfalls of different interpretations of data by different people in the organization and avoids guessing how severe a hazard is. Through safety policy, top management and the organization use the same risk management process and standard risk assessment table. Utilizing the standard risk management process and risk assessment table forces everyone in the organization, from top management to front-line employees, to view the hazard, the hazard analysis, the hazard assessment, the hazard mitigation, and hazard reevaluation from the same standard. While SMS enables the hazard risk to be evaluated more accurately, it also reevaluates the hazard more accurately after implementing mitigation strategies. It utilizes the same risk assessment table to lower the hazard risk after implementing mitigation controls. Suppose mitigation controls cannot appropriately manage the hazard within safe operational limits. In that case, the organization has the ethical responsibility to stop the operation until the hazard can be managed within safe operational limits. This scientific data-driven proactive approach to risk management is a vast improvement to safety programs of the past that used many different methods to manage risk along with different standards. SMS also provides a great deal of freedom to implement the risk management process by being flexible and allowing an organization to tailor the specific size and type of SMS for that aviation organization by tailoring the different SMS pillar elements. This type of customizable SMS safety program centered around risk management has allowed governing agencies like the FAA in the US to mandate and regulate SMS safety programs for different kinds of aviation organizations in the US where, in the past, safety programs were voluntary and cost-prohibitive. Being a proactive data-driven approach means that SMS is designed in each organization to identify hazards proactively through different types of data collection devices. In the past, aviation safety relied on reactive data to track incidents and investigate accidents. Then, after a thorough analysis, it identified the hazards and made recommendations to fix them. Unfortunately, in this reactive process, economic affordability could get in the way of prioritizing the hazards, and delays in mitigating many hazards could take years to resolve. With SMS, an airline can continuously monitor and analyze recorded flight parameter data from each revenue flight through a Flight Operations Quality Assurance (FOQA) program for the safety of flight operations. This does two significant things in fulfilling the data-driven SMS role. It first can identify individual crew mistakes or flight safety risks that, if deemed a hazard by the airline's FOQA committee, could be immediately mitigated through the SMS risk management process. More importantly, the tremendous amount of data being collected and analyzed over time could be instrumental in identifying a hazardous trend experienced by many crews and revenue flights within that airline and possibly outside that airline as well, and would also call for the SMS risk management and mitigation process to be utilized.

7. Integrating human factors into SMS

While SMS and Risk Management have proliferated throughout the aviation industry and have become common in the US aviation industry, human error still plays a direct role in aviation in upwards to 80% of the commercial accidents in the US. With human error playing such a major role in affecting aviation safety, aviation human factors are also starting to play a significant role in combating human error in aviation. We are currently at a juncture in aviation history where human factors and risk management are beginning to merge. Current accident and incident analysis clearly shows that multiple human factors are usually involved as hazards in triggering human error to cause accidents and incidents. The big question for aviation safety experts is how best to integrate human factors into SMS. This can be seen through three examples: in an operational setting on the flight deck with Threat Error Management (TEM) as an extension of Crew Resource Management (CRM), in a Fatigue Risk Management (FRM) System for Airline Maintenance and in the form of a PSAP/FOQA database of human factors root error causes to incidents.

As TEM becomes more commonplace on the flight deck of airlines as an extension of Human Factors and CRM, it could also be considered an SMS integration to the modern operational flight deck. In the case of CRM, 40 years of bringing human factors solutions to improve the crew working together on the flight deck with flight attendants, dispatch, maintenance, ATC is all through teamwork tenets like leadership, followership, task management, automation management, communications, assertiveness, situational awareness, and decision making. TEM is a proactive add-on to these previous CRM tenets in that it identifies threats or hazards to anticipate before the flight and briefs how they are to be handled, and how to handle errors quickly when they do occur during the flight before they get out of control. TEM can also help a crew recover from an unanticipated threat to the flight and be resilient. These threats and errors are hazards that have been identified and quickly analyzed. The only difference between TEM and an official SMS Risk Management system is that the threats and errors as hazards need to be thoroughly assessed with a risk assessment chart on the flight deck. However, the pilots then implement mitigation, reassess, and reevaluate the threats and errors as hazards. This could become a more accurate form of TEM embedded in the airlines' SMS by simply having the aircrews voluntarily add the threats and errors from each flight to an anonymous database. Pilot safety experts from that airline and potentially AI could then expertly assess and rate the threats and errors as hazards, along with how the threats and errors were mitigated during the flight. This data would make a complete loop of the current TEM model by keeping the ongoing model going in CRM and infusing that current TEM model with feedback from analyzed ongoing data to improve the airline's TEM.

FRM has been growing in the human factors method SMS-related method over the last two decades since aviation experts in the US identified fatigue in the early 1990s as a significant safety threat to aviation. As commercial pilots have had their crew rest regulations vastly improved over the last decade, the FAA has also encouraged airlines to use FRM programs. FRM currently can take an employee's fatigue assessment and rate that employee from a hazardous fatigue level to a low and acceptable level for work. Managers can also use it to recommend mitigation strategies to reduce fatigue levels in certain fatigued employees to make them acceptable for work. While FRM fits into the SMS program requirements quite nicely as far as Risk Management, Safety Assurance, Policy, and Promotion are concerned, the biggest challenge to making such systems commonplace is making them more accurate and more efficient to run, especially for other employee groups in the industry like Aviation Maintenance Technicians and Ramp workers that have very little in the form of rest regulations. The first problem with FRM systems is how to accurately measure fatigued workers with the multi-variable human factors that influence each individual's fatigue and how best to measure that person's fatigue through daily physiological measurement. The other issue beyond the challenge of the accuracy of the FRM is from a mitigation perspective for the personnel who have been identified as a fatigue hazard. Such a task could be quite challenging for any manager responsible for that personnel to manage that worker's fatigue within acceptable working levels. It is quite possible that AI human teaming with management could prove invaluable to both data collection and mitigation management of fatigued aviation personnel in the future if AI were to be able to be integrated into both the initial fatigue assessment and fatigue mitigation parts of the FRM process [9].

The ideal example of a proactive, data-driven, human factors integrated SMS solution would be a modern US Airline Pilot Safety Action Program (PSAP) combined with an airline's FOQA program. If every PSAP and FOQA event were broken down by the respective PSAP and FOQA committee members into a human factors root cause analysis, these root causes would greatly benefit the action of each committee in event mitigation accuracy for each crew involved. By rating each human factor's root cause involved in each event with the company's standard risk assessment chart, each human factor's root cause could then be sent into an anonymous company database. Over time, commonly identified human factors root cause hazards with the highest safety risk assessment could be identified as hazardous trends and dealt with by a special airline company committee to add the mitigation strategy for the hazardous trend. Once the mitigation strategy has been implemented, the data will continue to be monitored to ensure that the mitigation step is indeed working to lower the occurrence of the hazardous trend and that its risk is managed. The key to this example is having good data collection in the PSAP reporting system combined with the data analysis of the FOQA to form a strong and proactive event database. The human factors root cause analysis of each PSAP and FOQA event could then be expanded into the form of a human factors root cause database to identify hazardous trends. The human factors root cause database serves the SMS and SMS risk management process to proactively identify, analyze, and mitigate flight hazardous trends.

The primary goal for identifying human factors elements influencing SMS objectives is to recognize what issues invite or result in human or system error and how performance is affected by error, either more likely or less so. Human factors principles and elements are prominent in SMS, FRM, and most risk management and assessment efforts. Understanding that human factors entail a wide range of settings and performance is important. The ergonomics aspects, anthropometrics, measures, cognitive processing, automation interfaces, and many other considerations come under the human factors umbrella. Typically, these factors are dynamic, subject to individual variations and permutations, and often difficult to predict or calculate. For these reasons, it is sound practice to include a human factors specialist in the risk assessment planning and procedures.

8. Proactive risk analysis with XAI

Explainable Artificial Intelligence (XAI): a method that helps humans understand how outputs are generated by machine or deep learning algorithms. It quantifies model correctness, fairness, and transparency, enhancing artificial intelligence [AI]-assisted decision-making.

Integrating artificial intelligence (AI) and explainable AI (XAI) in aviation risk assessment has transformed safety and managerial procedures. Proactive Risk Analysis with Explainable Artificial Intelligence [XAI] is a ground-breaking method that blends advanced AI methods with the need for transparency and interpretability in safety-critical practices. This approach seeks to predict and mitigate risks ahead of occurrences or mishaps. It symbolizes a radical advance in aviation safety, leveraging the capability of AI while keeping the severe portion of human control and recognizing safety-critical resolutions.

As we can see in **Figure 3**, Explainable Artificial Intelligence (XAI) is at the heart of operations. XAI employs methods and techniques to generate clear explanations about how the AI operates and the rationale behind its decisions, thereby fostering human trust. XAI in aviation provides transparent risk assessments with clear justifications.

This approach has many advantages, from enhanced predictive capabilities in identifying delicate patterns and potential risks [10] to conducting real-time analysis, enabling proactive risk management [11]. This approach improves decision-making, comprehensive risk assessments [12], and increases efficiency. It enables standardization of different safety protocols across the industry [13].

This advanced approach encompasses multiple stages to ensure comprehensive risk management, starting with data collection, where aviation-specific data may be collected from different sources [12]. Advanced algorithms through AI/Machine learning techniques are then used to analyze the data and identify potential risks [10]. XAI techniques such as decision trees or rules-based explanations are utilized within the XAI layer [14, 15]. In stage four, the AI systems flag possible risks or uncommon models within operations. Aviation safety experts evaluate the AI's results and XAI-generated explanations, start the risk assessment evaluation, and craft mitigation strategies. Lastly, continuous monitoring ensures the effectiveness of the strategies as they keep refining the predictions.

Figure 3.
XAI risk assessment.

This approach presents challenges that require ongoing attention and refinement. It depends on the quality of the data and a significant data infrastructure. AI and predictive models are complex and need expertise in aviation and AI/ML; some stakeholders still have trust issues with the system [11, 14]. Moreover, there is potential bias [16] combined with high initial costs [17] and some legal and compliance issues [12].

More debate is needed on the potential influence of AI on risk assessment practices. One fact, though, remains constant, and that is the willingness of management to accept and implement the outcome recommendations of XAI. Delivering a more incisive and comprehensive evaluation will confront management with the same dilemma that has persevered for decades – disrupting established patterns, added cost, and reliable implementation of identified changes necessary. Acting upon XAI resolutions is still within the province of management liabilities and, consequently, a separate risk category altogether.

Through applications with Safety Management Systems models and processes, an area of rapid growth is using AI and machine learning to capture risk-related elements in several arenas like organizational practices, operator performance, and adherence to engineered standards. The computational statistics, the basis for machine learning, are regarded as not representational for general human intelligence or capable of human cognition operations. The primary interest is prediction in the decision-making process derived from data analysis [18]. As data aggregation methods and filtering processes become more available, organizations can experience substantial upgrades in risk management analyses. Some aircraft systems upload information regularly, which will be parsed into the appropriate elements of the risk assessment algorithms.

A factor related to trust in AI decisions is boundary conditions for the perceptions generated. Also, fairness in decisions generated by AI has been studied [19] to determine effects where recommendations by algorithms are preferred over those by humans, known as algorithmic appreciation. Generally, this occurs when AI decisions are compared with human expert decisions, although the same does not hold true when non-experts are involved. An added feature is the perception of risk associated with AI decisions, which makes human decisions more likely to take these considerations into account.

9. Conclusions

In conclusion, this chapter has offered significant insights. Large-scale decision-making is essential for daily activities in aviation operations and risk assessment. Since these decisions are made within impenetrable systems, people must understand how specific decisions are reached. Additionally, policymakers in aviation operations are increasingly concerned about the loss of explainability in AI-based systems models, which is a primary issue affecting the acceptability of AI-based decisions.

SMS is having an incredible impact in allowing the accuracy of risk management to be applied in the industry. The industry will become safer and more efficient as human factors integrate more into risk management and SMS to address the high percentage of human error concerns. Meanwhile, implementing AI into various SMS risk management systems looks promising in increasing data gathering and analysis while vastly improving the human-machine team's decision-making capability.

Explainable AI (XAI) plays a crucial part in aviation operations. Understanding the mathematical foundations of existing machine learning architectures may offer insights into how and why a result was obtained but not into the inner workings of the models themselves. This is essential for fostering trust and confidence in AI models within organizations. Moreover, AI explainability allows organizations to adopt a responsible approach to AI development.

Conflict of interest

The authors declare no conflict of interest.

Author details

Leila Halawi*, Mark Miller and Sam Holley
Embry Riddle Aeronautical University, Daytona Beach, Florida, USA

*Address all correspondence to: halawil@erau.edu

References

[1] Ericson CA. Hazard Analysis Techniques for System Safety. 2nd ed. Wiley; 20 July 2015. p. 640. ISBN-10: 1118940385. ISBN-13: 978-1118940389

[2] ICAO Safety Management Manual Doc 9859. Available from: https://skybrary.aero/sites/default/files/bookshelf/5863.pdf

[3] IATA Integrated Risk and Resilience Management Manual (IRRM). Available from: https://www.iata.org/en/publications/store/integrated-risk-resilience-management-manual/

[4] US Department of Transportation Federal Aviation Administration (FAA) Flight Standard Service. Risk Management Handbook. 2022. Available from:https://www.faa.gov/sites/faa.gov/files/202206/risk_management_handbook_2A.pdf

[5] ESARR4 Risk Assessment and Management in ATM (Eurocontrol). 2001. Available from: https://www.eurocontrol.int/sites/default/files/2019-06/esarr4-e10.pdf

[6] FAA Airport Cooperative Research Program- ACRP Report 50. Improved Models for Risk Assessment of Runway Safety Areas. Available from: https://www.icao.int/sam/documents/2011/agaaserostudies/acrp%2004-08%2050-draft%20final%20report.pdf

[7] FAA Office of Airports Safety Management Systems. 2023. Available from: https://www.faa.gov/sites/faa.gov/files/ARP-SMS-Desk-Ref-v2-0.pdf

[8] ISO 31000. Risk Management – Guidelines, International Standard. Geneva: International Organization for Standardization; 2018

[9] Miller M, Mrusek B, Herbic J, Holley S, Halawi L. Implementing an AI fatigue risk management system for aviation maintenance SMS: A technology enhanced critical process human factors safety plan. In: Proceedings of the AHFE. Honolulu, Hawaii; 2024. Available from: https://www.hawaii.ahfe.org/program.html

[10] Degas A et al. A survey on artificial intelligence (AI) and eXplainable AI in air traffic management: Current trends and development with future research trajectory. Applied Sciences. 2022;**12**(3):1295. DOI: 10.3390/app12031295

[11] Ayhan S, Pesce J, Comitz P, Sweet D, Bliesner S, Gerberick G. Predictive analytics with aviation big data. In: 2013 Integrated Communications, Navigation and Surveillance Conference (ICNS), Herndon, VA, USA. 2013. pp. 1-13. DOI: 10.1109/ICNSurv.2013.6548556

[12] Kistan T, Gardi A, Sabatini R. Machine learning and cognitive ergonomics in air traffic management: Recent developments and considerations for certification. Aerospace. 2018;**5**(4):103. ProQuest. Available from: https://www.proquest.com/scholarly-journals/machine-learning-cognitive-ergonomics-air-traffic/docview/2582792906/se-2. DOI: 10.3390/aerospace5040103

[13] Christine B et al. Hazards identification and analysis for unmanned aircraft system operations. In: 17th AIAA Aviation Technology, Integration, and Operations Conference. Denver, Colorado; 5-9 June 2017. DOI: 10.2514/6.2017-3269. Available

from: https://arc.aiaa.org/doi/book/10.2514/MATIO17

[14] Ribeiro TM, Singh S, Guestrin C. 'Why should I trust you?': Explaining the predictions of any classifier. In: Proceedings of the 22nd ACM SIGKDD International Conference on Knowledge Discovery and Data Mining. New York, NY, USA: ACM; 2016. pp. 1135-1144

[15] Lundberg, Scott, and Su-In Lee. "A Unified Approach to Interpreting Model Predictions." arXiv.Org. 2017. DOI: 10.48550/arxiv.1705.07874

[16] Mehrabi N et al. A survey on bias and fairness in machine learning. ACM Computing Surveys. 2021;**54**(6):1-35. DOI: 10.1145/3457607

[17] Gürbüz F, Özbakır L, Yapıcı H. Explainable artificial intelligence for aviation safety: A review. Expert Systems with Applications. 2023;**213**:118877

[18] Agrawal A et al. Exploring the impact of artificial intelligence: Prediction versus judgment. Information Economics and Policy. 2019;**47**:1-6. DOI: 10.1016/j.infoecopol.2019.05.001

[19] Araujo T et al. In AI we trust? Perceptions about automated decision-making by artificial intelligence. AI and Society. 2020;**35**(3):611-623. DOI: 10.1007/s00146-019-00931-w

Chapter 5

Extending the Range of Electric Aircraft Using Infrastructure Objects

Sergej Težak and Drago Sever

Abstract

This chapter presents the concept of infrastructure objects that can be used to increase the range of electric aircraft. First, the achievements of electric aircraft powered by batteries will be introduced. The main issue lies in their limited range and the considerable weight of the batteries. To enhance this, the chapter presents infrastructure solutions enabling electric aircraft to fly longer distances. Conventional infrastructure solutions are depicted as expanded airports with additional parking spaces equipped for recharging aircraft with electric power, as well as in the form of floating airports for transoceanic flights. Unconventional infrastructure solutions are portrayed as uplifting towers that generate sufficient updrafts to provide extra altitude for electric aircraft. Simultaneously, these towers can function as power stations harnessing energy from the updraft winds. Based on this, a flying system can be established for aircraft that require minimal or even no additional energy for flight.

Keywords: electric aircraft, flight range, battery recharge, updraft tower, air transport, glider, small aircraft

1. Introduction

Due to increasing environmental pollution, humanity has decided to adopt modes of transportation that have less impact on the environment. Existing transportation means have a negative impact on the environment mainly through emissions of harmful gases and noise [1]. After their service life, they also pollute the environment as waste material that cannot be recycled. Therefore, in the last decade, electric mobility has been increasingly coming to the forefront, particularly in road transport. In air transport, individual electric aircraft are beginning to appear, but the use of these aircraft is currently very small. In the case of introducing electric vehicles into traffic, the development of road vehicles currently exceeds the development of aircraft. In history, it was always the opposite and the aviation industry was considered the flagship in the introduction of technical innovations, which were later adopted by road vehicles. The problem with aircraft powered by electricity is that the batteries are still too heavy in terms of energy value for use in air traffic [2]. Therefore, the range of electric aircraft on batteries is short compared to aircraft with conventional propulsion with internal combustion engines or turbine engines [3]. However, much is expected from aircraft with hydrogen fuel cells [4]. However, even in this case, the problem is with weight and also storing and supplying hydrogen to the vehicle still presents mostly safety problems. The cost of these systems is also more expensive for the time being.

The problem with electric-powered airplanes is also their lower speed, as current and planned airplanes are primarily propeller-driven, resulting in lower speeds compared to turbine-powered jets. However, in passenger air travel, cost is the primary concern, with the goal being to make air travel more affordable for passengers. This is why fast super-sonic passenger planes, such as the Concorde and Tu-144, became noncompetitive and are no longer in use, as their fuel consumption was simply too high. In the future, this may no longer be a problem, as passengers may be willing to spend more time traveling as long as they have sufficient comfort and can work or engage in other leisure activities during the flight. This primarily refers to access to communication (Internet), watching movies or television programs, access to food and toilets, and, of course, sufficient space for each passenger during the flight to engage in these activities.

In the last years, the COVID-19 pandemic has completely disrupted air travel. Large planes are particularly problematic when it comes to infections. If the danger to people does not subside in the future, smaller vehicles will come to the forefront of passenger transportation, including airplanes. Electric planes, in particular, could play a crucial role here, as smaller planes are expected to be used, where there is no risk of disease transmission. In this regard, the most optimal way to travel would be within the individual family or "bubble," which means small planes with 4 to 9 passengers, which is also the focus of research and development in the case of electric planes [5].

Since small planes also have a limited range, it would be necessary to create a dense network of airports with enough space for battery charging. Smaller airports located close to the flight paths of the planes would also be suitable, even if they are not located in larger cities.

Ways to increase the range of planes with infrastructure have not yet been fully explored. This primarily involves finding ways to increase the potential energy of the plane with the help of thermal updrafts, which are already used by glider planes.

2. Methodology

The proposed methodology is mainly based on gathering data from previous research in various fields that could be relevant to solving the problem of increasing the range of electric planes powered by batteries. Solutions used in other disciplines and activities could be useful in addressing this problem. This is primarily the focus of the TRIZ method and a similar method called IN-IN, developed by the aircraft designer Bartini [6]. These methods solve the problem by exploring solutions in other disciplines or determining whether similar solutions have been used to solve the current task or problem.

To solve the problem, computational models will be used based on 3D models of infrastructure objects. First, geometric models will be presented, which will contain new ideas, and based on these models, established computational models will be used to demonstrate the possibilities of using infrastructure objects to increase the range of small electric planes powered by batteries.

3. Current state of development and use of electric aircraft with batteries

In this chapter, we will present the electric aircraft that are already in use, as well as some developmental projects for electric aircraft that are currently underway and will be available for use in the coming years. The problem with introducing passenger aircraft into service is primarily the fact that it takes much longer to certify them than it does for the automotive industry. Materials, components, and individual subsystems of aircraft are subject to strict controls, primarily from a safety perspective, in aviation. Therefore, some aircraft require several years from the time they are developed until they are brought to market.

The data obtained for electric planes is constantly changing, as most planes are still in prototype phase. First certified and available on the market was the Pipistrel Velis Electro [7]. **Table 1** shows the top characteristics of today's electric planes. Smaller electric planes with two passengers can travel about 110 nm (200 km). This applies to smaller two-passenger planes. Larger planes have a greater range. The longest-range electric plane currently in development is the Eviation Alice [8]. The projected range is 440 nm (815 km), which is already a high number. However, in practice, this range is reduced to 290 nm + 30 minutes reserve. This is still four times less than comparable turboprop planes. But the most interesting aspect is the cost of flying, which is only $200 per flight hour compared to turboprop planes that have costs between $1200 and $2000 (in the year 2022). The cost of a new battery would be approximately $300,000 (in the year 2023). Electric planes are made of lightweight composite materials so that the remaining weight can be used for the weight of the batteries. The Alice plane has a 3720 kg battery. The remaining parts of the plane, including the motor, weigh only 1400 kg. The weight of the cargo is 1100 kg. For comparison, the empty weight of a turboprop plane Pilatus PC-12 is 2810 kg, which is twice the weight of an empty electric plane without batteries. From these data, it is clear how electric plane constructors must use the lightest materials in aircraft construction. However, the data in the table show that electric planes also achieve extremely fast range development. In 2015, the developed Pipistrel Alpha Electro had a range of only 160 km, which is approximately 7 times less than the conventional Cessna Skyhawk [1]. The range of the Pipistrel Panthera electro and the eFlyer 2 electric planes is 400 km, which is only 3 times less than that of a comparable conventional plane. eFlyer 2 has a high gliding ratio of 20.6:1. Its batteries allow for 3 hours of flight and half an hour of reserve [10]. Today's batteries allow for a maximum energy of 220 Wh/kg to 250 Wh/

Aircraft	**Max. range (nm / km)**	**Passengers (Nr.)**	**Max.cruise speed (knots / km/h)**	**Empty Weight (kg)**	**Power (kW)**	**Battery (kWh)**
Bye Aerospace eFlyer 2	220/400	2	135/250	662	110	?
Pipistrel Velis Electro	108/200	2	90/180	428	57,6	24,8
NASA X-57 Maxwell	100/180	4 ?	135/250	1400	2 x 60	47
Pipistrel Panthera elec.	220/400	4	175/325	?	200	?
Eviation Alice	440/815*	2 + 9	250/463	5200 ?	2 x 634*	820*

?—The data are not available or are only an estimate.
**—Projected data.*

Table 1.
Electrical aircrafts [7–11].

kg, but in the future, it is expected that they could reach up to 500 Wh/kg. With the development of batteries, we can also expect an increase in the range of electric planes of up to 100% [12].

4. Supporting infrastructure for increasing the range of electric planes

Since electric planes have a limited range, careful planning and construction of supporting infrastructure will be necessary. This includes airports and other supporting facilities.

4.1 Conventional infrastructure approach to support electric planes

In the conventional sense, this means that on longer flight routes, planes will land at intermediate airports where they can recharge their batteries. Since charging will take up to half an hour, airports should be designed in such a way that planes do not waste too much time taxiing to the charging station (**Figure 1**). Since some routes will be heavily loaded, it would make sense to build multiple airports in the direction of the flight path and for airports to have multiple landing strips with charging stations placed as close as possible. This way, the plane would lose no more than 45 minutes at each airport for landing, taxiing to the parking lot, recharging batteries, taxiing to the takeoff point, and taking off.

The biggest challenge, however, is certainly flying over oceans, intercontinental flights. In this case, airports would have to be built on islands or floating airports. The largest floating structure today is the Prelude FLNG [13], which is 488 meters (1601 ft) long, 74 meters (243 ft) wide, and weighs 260,000 tons. This length is too short for takeoff and landing of the Aviation Alice aircraft. However, there are already systems available that assist in aircraft takeoff and landing, such as those used on aircraft carriers. On the other hand, modular floating airports are also being developed. The technology for this already exists [14]. The idea of floating airports is not new and was first proposed by Geoffrey Pike during World War II. His idea was to build large floating airports out of pykrete (a mixture of ice and sawdust) that he

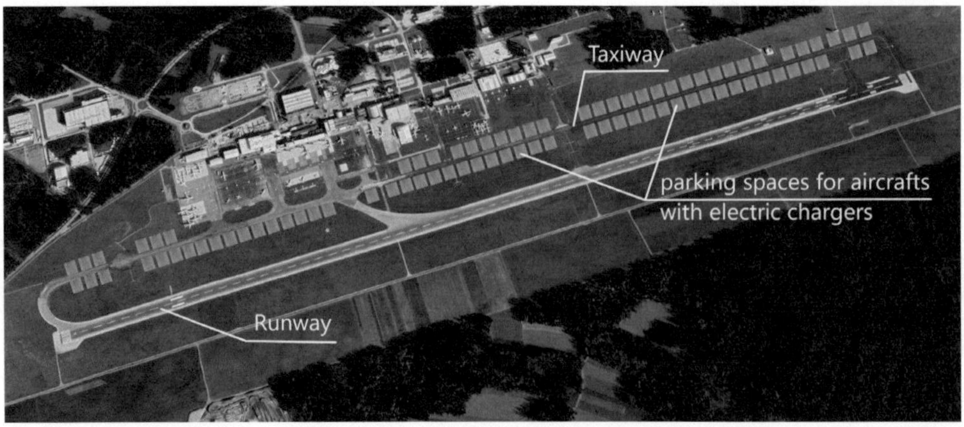

Figure 1.
Airport with additional 85 parking spaces with electric chargers for electric planes along the taxiway (Airport Ljubljana).

had developed, on which planes would land in the Atlantic Ocean. This material is lighter than water and has very good mechanical [15]. In this way, convoys of cargo ships could be protected from dangerous German submarines. The project was called Habbakuk, but it was never realized [16]. They only made a pykrete ship and demonstrated the excellent properties of this material.

In 2000, a test floating airport was built in Tokyo Bay, Japan. The runway was 1000 meters long and was successfully tested with aircraft landings. The airport was dismantled after six years [17].

One of the challenges for floating airports would be parking spaces for charging electric airplanes. They would need a large footprint because many planes would fly in the same direction on an oceanic route, taking off and landing at the airport. Therefore, they would require numerous parking spaces. Another solution would be for planes to go to lower floors under the runway where electric charging stations would be located (**Figure 2**). This way, they would also be protected from the severe weather conditions that can occur on the ocean. As a significant amount of energy would be required for charging the planes, nuclear energy, which is already being used on aircraft carriers, would be suitable for these floating airports.

The airport would be composed of individual modules, each around 400 meters long, which would be transported separately to the location where the floating airport would be erected (**Figure 3**).

These airports should be located within a range of 500 km to 600 km if electric planes have a range of 800 km. It is better for them to be closer together because the weather over the ocean can be very stormy, and in such cases, headwinds can significantly shorten the planes' range. With the development of more powerful batteries, the distances between floating airports would increase, and fewer floating airports would be needed on each flight route. Excess floating airports would be moved to new routes. In the case of a flight from London to New York, five floating airports would be needed (**Figure 4**).

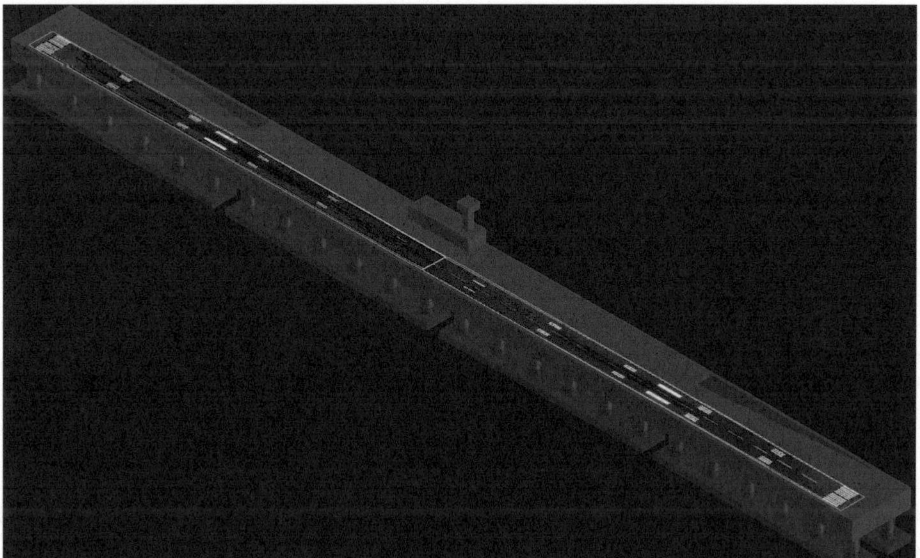

Figure 2.
Floating airport.

Figure 3.
Individual modules of the floating airport.

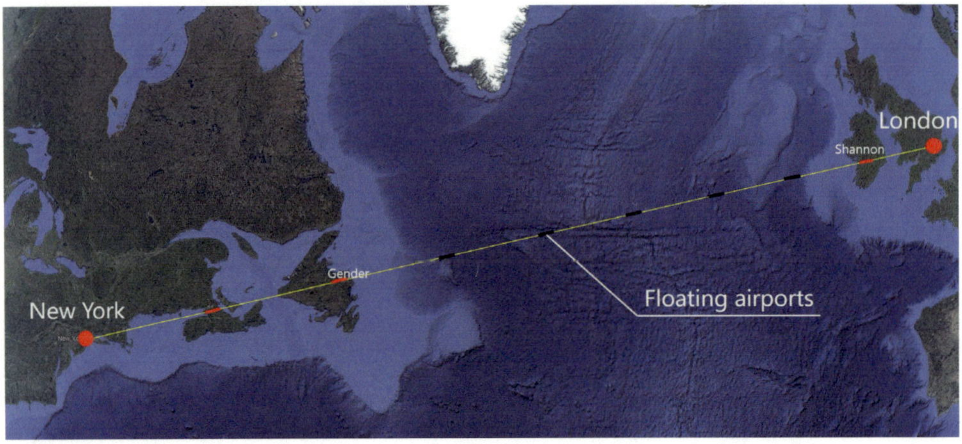

Figure 4.
A flight from London to New York via *five floating airports.*

Airports on land located along the flight route (such as Shannon, Gender ...) have already been built; they would only need to be equipped with infrastructure for electric planes. It would probably be necessary to move the flight route further south, as there are icebergs in the vicinity of Newfoundland that could be dangerous for floating airports. This would lengthen the flight and require more floating airports.

4.2 Unconventional infrastructure approach to support electric planes

The idea of infrastructure that literally increases the range of aircraft originated from the field of gliding. Thermal lift is often used in glider flying, in which air rises, allowing the glider to increase its potential energy, which it later converts into kinetic energy during flight. Thermal lift occurs in sunny weather due to various factors, most often due to the difference in heating of the ground by sunlight. This causes the air to warm up more in some places than in others. As the warm air rises, cold air from the surrounding areas takes its place, which is then warmed up by the ground. However, sometimes glider pilots arrive at the airport very late, when the sun is already setting. In conversation, they revealed their secret that the thermal lift above the aluminum factory always works in the evening. They say that they could glide there all night (A conversation with the glider pilot Iztok Žagar in 2001) The fact is that the aluminum factory has extremely powerful furnaces for melting, which emit a huge amount of heat into the environment, warming up the air near the ground. And this air rises even when there is no sun. In windy weather, the thermal lift can tilt or be interrupted by strong winds. Then, glider pilots have a hard time finding rising air. The problem is that thermal lift is not a permanent phenomenon and it is not possible to predict exactly when and where it will occur. Because we do not yet know of glasses or cameras that can see warmer, rising air at a distance of several kilometers, glider pilots at that time have difficulty finding rising air. Experience (thermal lift is often under cumulus clouds) and also observing birds, which have natural senses to find rising air, often help the most. In fact, birds have instincts, experience, and knowledge with which they can fly with the least energy consumption. Albatrosses are able to fly long distances without flapping their wings. Years ago, scientists even discovered the way they fly, enabling them to fly over oceans [18]. Termites also know a similar way of wandering, building high termite mounds in which the air is heated and rises. In this way, termites quickly reach a high altitude and fly long distances when they take off.

Based on the example from nature, we could also build towers in which air would be heated and lifted, allowing airplanes to circle within these towers and gain altitude. An important factor is the height of the towers. Cooling towers for nuclear or thermal power plants are already known to reach heights of up to 210 meters and a diameter of 174 meters [19]. However, towers in which airplanes could circle would have to be much larger. The tallest building today is the Burj Khalifa, which is 828 meters high. However, buildings over 1000 meters are also being planned (Burj Mubarak al-Kabir). In science, it is already predicted that it will be possible to build buildings up to 1500 meters high [20]. This means that in the future, it will be possible to build towers with a diameter of 1.5 kilometers and a height of 1.5 kilometers (**Figure 5**). The diameter must be large enough for airplanes with up to 12 passengers to circle inside the tower and gain altitude.

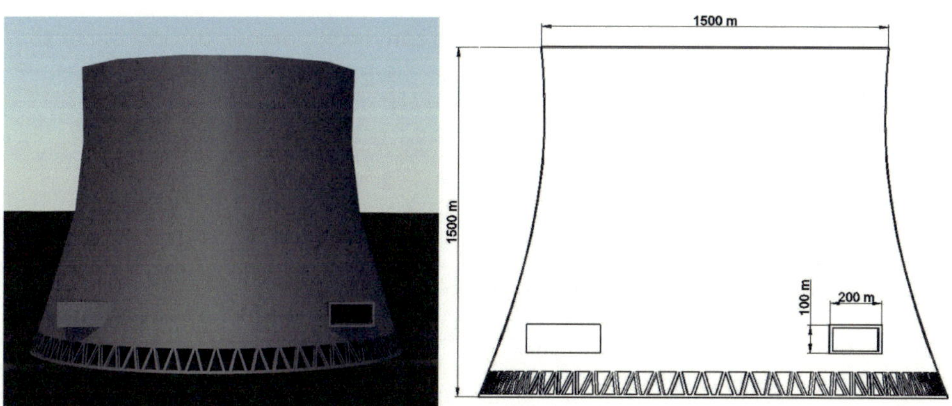

Figure 5.
Sketch of a lifting tower with dimensions.

The question is what the speed of air lift would be. In the calculation of natural lift, data from the standard atmosphere is used, which means that the air is at an altitude of 0 meters, has a temperature of 15°C, a pressure of 11,325 Pa, and a density of 1225 kg/m^3. In the troposphere, temperature and pressure decrease with altitude. The temperature gradient of decreasing temperature with altitude is 6,5°C/km (lapse rate) in the troposphere. The drop in pressure in the tower depends on the change in temperature between the air in the tower or the incoming air and the temperature of the surrounding air at the tower's exit, as well as the height of the tower. If we consider that the air is an ideal gas, the pressure difference in the tower can be calculated using the following [21]:

$$\Delta p = g \cdot \rho_D \cdot \left(\frac{T_D - T_O}{T_O} \right) \cdot H \quad (1)$$

Δp—pressure drop (Pa).
ρ_D—density of the air at the entrance (kg/m^3).
g—gravitational acceleration (m/s^2).
T_D—temperature of the air at the bottom of the tower, at the entrance (K).
T_O—temperature of the air at the exit of the tower (K).
H—height of the tower (m).

Assuming that the tower is located at an altitude of 0 m and that standard conditions (standard atmosphere) apply at the tower's entrance at the bottom, then the air temperature at the entrance T_D = 15°C. The same applies to pressure and air density. Since air temperature in lower layers (troposphere) decreases on average by 6,5°C/km with height, the temperature of the surroundings at the tower's exit is as follows:

$$T_O = T_D - H \cdot \frac{(dT / dz)}{1000} = 15 - 1500 \cdot \frac{6,5}{1000} = 5,25^O C \quad (2)$$

(dT / dz)—temperature gradient (6,5°C/km).

The temperature of the ambient air at a height of 1500 m is 5,25°C or 278,25 K. The pressure drop through the chimney will be as follows:

$$\Delta p = g \cdot \rho_D \cdot \left(\frac{T_D - T_O}{T_O} \right) \cdot H = 9,81 \cdot 1,225 \cdot \left(\frac{288 - 278,25}{278,25} \right) = 631,63 Pa \quad (3)$$

Some of the potential energy in the tower is converted into kinetic energy of the air flowing through the tower, while the rest is used to overcome friction. The wind speed is determined by the following equation [21]:

$$\Delta p = \left(1 + \lambda \cdot \frac{H}{2 \cdot r} \right) \cdot \frac{\rho_D \cdot V_D^2}{2} \quad (4)$$

r—tower radius (m).
V_D—velocity of rising air (m/s).
Since the tower is very wide, friction can be neglected.

$$\Delta p = \frac{\rho_D \cdot V_D^2}{2} \quad (5)$$

$$V_D = \sqrt{\frac{2 \cdot \Delta p}{\rho_D}} = \sqrt{\frac{2 \cdot 631,63}{1,225}} = 32,11 m / s \quad (6)$$

The height of the tower provides a high value of updraft wind, up to 32 m/s. This speed can be reduced by closing off the air supply at the bottom entrance. Since the speed and temperature of the air at the exit would still be quite high, the air would rise even higher above the tower and could provide updraft wind to planes at higher altitudes, even up to 5 km. Such towers could also be used as updraft power plants, operating on the same principle as solar updraft power plants. The updraft wind would be sufficient to power turbines with generators at the tower entrance (**Figure 6**).

With the electricity generated, electric planes landing at nearby airports could be charged. When there is no natural updraft (during inversion), power turbines can be used as propulsion fans to create necessary updraft wind. As electric planes consume a lot of energy to climb to cruising altitude, these updraft towers could save energy. If an airplane has a glide ratio of 20.6:1, like the eFlyer 2, it could increase its flight range by at least 100 km from a height of 5 km, and also have fully charged batteries at that altitude, further increasing its flight range. As this airplane has a flight range of 400 km, the updraft towers could be spaced at intervals of 500 km. For larger airplanes such as the Eviation Alice, the updraft towers could be spaced at intervals of 1000 km. However, during the time planes circle inside the tower, they would need to use their electric motors as generators to charge their batteries. If the air rises at 30 m/s, the plane can relatively descend in this rising air at 25 m/s while still ascending at an average speed of 5 m/s. While the plane is relatively descending in this rising air, the propellers would act as wind turbines and charge the batteries. They would also act as brakes. If the plane ascends at an average speed of 5 m/s, it would take 16.6 minutes to reach a height of 5 km. In this time, the generators would need to charge the batteries. However, it is questionable whether the electric motors that would act as electric generators have enough power. If the electric motors in Eviation Alice worked as generators with a full power of 2 x 634 kW, it would take 38.8 minutes to fully charge the 820 kWh battery. Alternatively, if one wanted to charge the battery

Figure 6.
Updraft towers with wind turbines that can function as power plants to generate energy or as propulsion fans to create updraft wind during inversion (when there is no natural updraft).

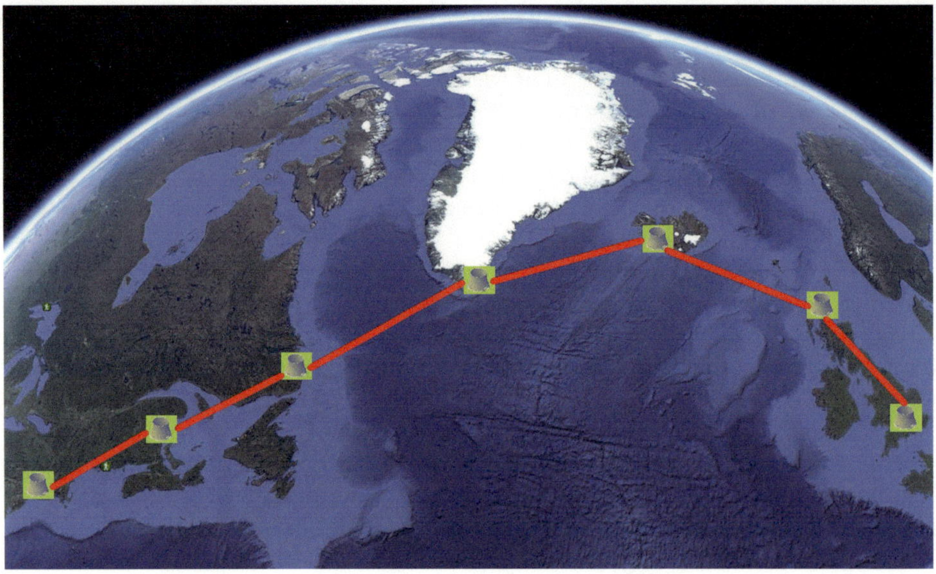

Figure 7.
Travel plan from London to New York with land-based lifting towers.

faster, the aircraft would need additional air turbines (wind turbines) that would be pulled out of the fuselage during lifting to charge the battery. For an 820 kWh battery, a total power of 3000 kW would be required from the air turbines (wind turbines) to fully charge the battery in 16.6 minutes. For this purpose, airplanes would need additional aerodynamic elements (flaps, retractable elements) that would reduce the aircraft speed to a minimum and allow it to circulate at a lower speed within a tower.

By arranging such large towers at key points, airplanes that would have increased range up to 1200 km could travel the distance from Europe to the US beyond the already known air routes used by smaller airplanes (**Figure 7**). The route runs from the United Kingdom, Iceland, Greenland, to the American continent. However, the problem of ocean flights would be a currently unsolvable problem because it is difficult to build such large towers that would float. The stability issue would be the most critical concern during strong winds. Most of such an object would be underwater, so it would be best to consider building such towers where the ocean depths are shallower and to create artificial islands. This method of flying can be combined with conventional floating airports, as described in chapter 4.1. The time it would take to travel from London to New York (*via* Iceland and Greenland) covering a distance of approximately 6130 km, for an aircraft with around 12 passengers (similar to the Eviation Alice) and a cruising speed of 250 nm (463 km/h) would be 14 hours for the flight and about 100 minutes for ascent in the lifting towers, totaling 15 hours and 20 minutes. This is how much time airplanes needed for the same route in the 1950s.

5. Flying without adding energy

Here is an idea presented on how people could fly without adding energy to the airplane or with airplanes requiring only minimal energy to reach the first updraft tower and increase their altitude. Smaller towns would have smaller updraft towers up to 500 m high, while larger cities would have larger updraft towers over 1.5 km high. Gliding planes for one family (4 to 6 people) would be used, which would have good gliding abilities. Nowadays, top of the line gliding planes already have a gliding ratio of 50:1 and even up to 70:1 (Eta glider). However, the problem is that they achieve such top results at lower speeds (108 km/h). They also have a large wingspan of 30 m. For long distance flights, gliding planes for 4 to 6 people that provide full passenger comfort (including a restroom) would have to be developed. It would need to have a gliding ratio of 50:1 at the highest speed (e.g., 200 km/h). It should have a reserve of electrical energy for at least 200 km of flight, in case of strong headwinds. In this case, large updraft towers that would enable airplanes to reach an altitude of 5 km would have to be placed at a maximum distance of 200 km from each other—near large cities (**Figure 8**).

If the entire planet were built with such towers, people could travel all around the world. However, travel would take a long time, more than one or two days. The gliding plane could take full advantage of the updraft column and would rise at speeds of up to 15 m/s or more. This way, for example, such a plane would glide from Zagreb to Phuket, a distance of 8780 km, in less than two days (44 hours) at an average speed of 200 km/h. In between, it would have to ascend 43 times if the updraft towers were 200 km apart. In these towers, it would take an additional 4.3 hours to lift the plane to its maximum altitude. The whole journey would take 48.3 hours, or two days. Therefore, the airplane would have to provide maximum possible comfort for passengers, including the possibility of sleeping. The airplane would drive automatically and would not consume any energy for takeoff. Once these towers are built, future generations could still use them. Good gliding planes will have to be produced to make this possible. This way, travel would be almost free. The airplane, if mass produced, would be worth as much as a car today. They would sell a lot, especially with the installation of solar cells on the plane's wings, which would even increase the plane's range or speed on sunny days.

The advantage of such a mode of travel is that it would decentralize air traffic. In fact, we would only need to travel from home to the first grassy airport, which is no more than 10 km away. From there, we would fly directly to the first major updraft tower and then on to our destination *via* intermediate updraft towers, to a small airport that is very close to our travel destination. Currently, when we travel far, we first have to travel by car (taxi), train, or bus from home to the nearest regional airport, which can be up to 100 km away (e.g., Zagreb). Then, from there, we fly to

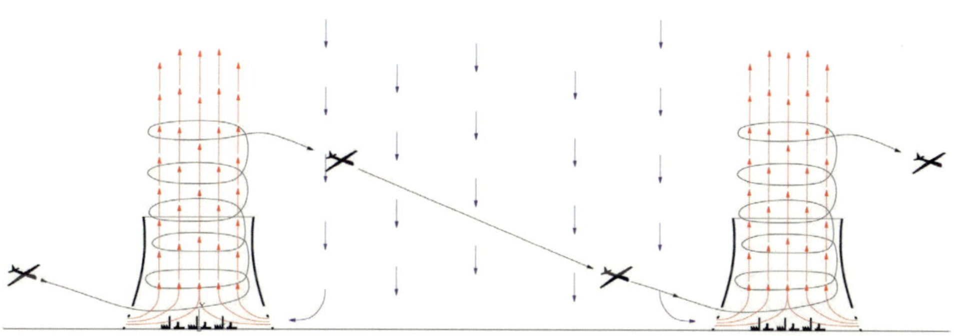

Figure 8.
The flight from one updraft tower to the next updraft tower.

the first major hub airport (e.g., Istanbul) and transfer to a larger plane. Changing planes at these airports can be very stressful because of the crowds, and we are afraid of missing our next flight. Then, we fly on a larger plane to another major international airport near our destination (e.g., Bangkok). There, we have to go to another airport in Bangkok where local flights are conducted. We then board a smaller plane and travel to a local airport that is close to our destination (e.g., Phuket). From there, we take a taxi to our destination, which can be up to 40 km away. Such travel is very tiring and can take a day and a half, especially if we do not have immediate consecutive flights and have to wait at intermediate airports. Unless we live in larger cities, where there are direct flights to desired destinations.

6. Conclusion

Air passenger traffic is sure to switch to more environmentally friendly flying methods in the future. The technology of electric cars using batteries, which is already used in automobiles, is beginning to be used in aviation as well. However, the problem with airplanes is the large weight of the battery, which can be three times heavier than the weight of the airplane with electric motors without batteries. Therefore, the lightest materials, such as artificial plastics, must be used to build such aircraft. And still, the range of such planes is 3 to 4 times smaller than the range of comparable turboprop planes. The maximum range that electric battery-powered planes can achieve with today's technology is around 800 km. This range could be extended with infrastructure facilities.

In a conventional way, with the establishment of several airports along the aircraft's route, the airports would be spaced approximately 500 km apart. The airports should provide spaces for charging the planes with electricity as close to the runway as possible, to lose as little time as possible for taxiing the plane on the ground.

Alternatively, the range of planes could be increased with updraft towers that would be large enough for planes to circle inside and gain altitude (height of 1500 m and diameter of 1500 m). If the updraft tower is high enough and is combined with an updraft solar power plant, the updraft wind would be large enough for the plane's propeller generators to charge the batteries during the circling. To charge the batteries quickly enough, the plane would have to deploy additional wind turbines with generators while circling. This would increase the range of the planes, as they would reach an altitude of 5 km with full batteries and would only use energy for horizontal flight and landing. With such supportive infrastructure, planes could also have a range of up to 1200 km. In this way, aircraft with up to 12 passengers and a cruising speed of 463 km/h would need 15 hours and 20 minutes to travel from London to New York (if the route goes *via* Iceland and Greenland).

If these large updraft towers were placed densely enough at a distance of 200 km, close to larger cities, people could fly with glider planes without additional energy. Glider planes for 4 to 6 people with maximum travel comfort and an average flight speed of 200 km could offer energy-free travel. If they also had solar cells on their wings and a small battery for the electric motor, they could actually fly for free. Given the situation in aviation, where there has been a large drop in passenger traffic in the last two years, small planes with a small number of passengers are ideal for an epidemic. Planes with 4 to 6 passengers and up to 12 passengers will be the most successful in such circumstances, and the solutions for extending the range of such planes shown in this chapter would enable the breakthrough of long flights with small planes from small airports (glider planes can also use grassy airstrips). This would decentralize air transport and no longer require the largest collection airports, where passengers come with regional flights to fill large planes for transoceanic flights.

Author details

Sergej Težak* and Drago Sever
Faculty of Civil Engineering, Transportation Engineering and Architecture, University of Maribor, Maribor, Slovenia

*Address all correspondence to: sergej.tezak@um.si; sergej.tezak@gmail.com

References

[1] Thapa N, Ram N, Kumar S, Mehta J. All electric aircraft: A reality on its way. Materials Today: Proceedings. 2021;**43**:175-182. DOI: 10.1016/j.matpr.2020.11.611

[2] Donateo T, Ficarella A, Spedicato L, Arista A, Ferraro A. A new approach to calculating endurance in electric flight and comparing fuel cells and batteries. Applied Energy. 2017;**187**:807-819. DOI: 10.1016/j.apenergy.2016.11.100

[3] Justin CY, Payan AP, Briceo SI, German BJ, Mavris DN. Power optimized battery swap and recharge strategies for electric aircraft operations. Transportation Research Part C. 2020;**43**:102605. DOI: 10.1016/j.trc.2020.02.027

[4] Renau J, Lozano A, Barroso J, Miralles J, Martin J, Sanchez F, et al. Use of fuel cell stacks to achieve high altitudes in light unmanned aerial vehicles. International Journal of Hydrogen Energy. 2015;**40**:14573-14583. DOI: 10.1016/j.ijhydene.2015.02.071

[5] Grimme W, Paul A, Maertens S, van Wensveen J. The prospect of hybrid-electric regional air transport—An assessment of travel time benefits of domestic short-haul flights in Germany with 10-seater aircraft. Transport Research Procedia. 2020;**51**:199-207. DOI: 10.1016/j.trpro.2020.11.022

[6] Težak S. Red wings proposed by Robert Bartini for sustainable aviation. In: Agarwal R, editor. Environmental Impact of Aviation and Sustainable Solutions. London: IntechOpen; 2020. pp. 193-204. DOI: 10.5772/intechopen.85032

[7] Pipistrel, Velis electro. [Internet]. 2024. Available from: https://www.pipistrel-aircraft.com/products/velis-electro/

[8] Watch Eviation's all-electric Alice aircraft first flight. [Internet]. 2023. Available from: https://electrek.co/2022/09/27/eviation-all-electric-alice-aircraft-first-flight/

[9] BYE Aerospace, Electric training aircraft. [Internet]. 2024. Available from: https://electricflyer.com/

[10] Pipistrel. Panthera electric. [Internet]. 2024. Available from: http://www.ifly.eu/Panthera/Electric

[11] NASA X-57 Maxwell [Internet]. 2024. Available from: https://en.wikipedia.org/wiki/NASA_X-57_Maxwell#X-57_Maxwell

[12] CATL battery successfully powers electric plane with 1,800-mile civil aircraft expected. [Internet]. 2024. Available from: https://electrek.co/2024/06/25/catl-successfully-tests-electric-plane-1800-mile-model-nears/

[13] Shell's gargantuan 'Prelude' is the world's largest floating structure—And it just left for Australia. [Internet]. 2024. Available from: https://www.businessinsider.com/shells-gargantuan-prelude-is-the-worlds-largest-floating-structure-2017-7

[14] Jiang D, Tan KH, Dai J, Ang KK, Nguyen HP. Behavior of concrete modular multi-purpose floating structures. Ocean Engineering. 2021;**229**:108971. DOI: 10.1016/j.oceaneng.2021.108971

[15] Pronk A, Mistur M, Li Q, Liu X, Blok R, Liu R, et al. The 2017-18 design and construction of ice composite

structures in Harbin. Structure. 2019;**18**:117-127. DOI: 10.1016/j.istruc.2019.01.020

[16] Project Habbakuk: Britain's SecretIce "Bergship" aircraft carrier project. [Internet]. 2024. Available from: https://99percentinvisible.org/article/project-habbakuk-britains-secret-ice-bergship-aircraft-carrier-project/

[17] Wang CM, Tay ZY. Very large floating structures: Applications, research and development. Procedia Engineering. 2011;**14**:62-72. DOI: 10.1016/j.proeng.2011.07.007

[18] Richardson PL. How do albatrosses fly around the world without flapping their wings? Progress in Oceanography. 2011;**88**(1-4):46-58. DOI: 10.1016/j.pocean.2010.08.001

[19] Shitang K, Wenlin Y. Wind load characteristics and action mechanism on internal and external surfaces of super-large cooling towers under wind-rain combined effects. Mathematical Problems in Engineering. 2018;**2018**:22. DOI: 10.1155/2018/2921709

[20] Boretti A, Al-Zubaidy S. Maturity assessment of the solar updraft tower technology. Renewable Energy Focus. 2018;**27**:135-144. DOI: 10.1016/j.ref.2018.10.001

[21] Kuštrin I, Tuma M. Sončni dimnik (Solar chimney). Strojniški vestnik (Journal of Mechanical Engineering). 1985;**year31**(11/12):309-314

Chapter 6

Aerosol Jet-Printed Sensors for Aero-Engine Component Health Monitoring

Edgar Jeevan Danaraj

Abstract

Aerosol jet printing (AJP) is emerging as a versatile additive manufacturing technique, particularly suitable for fabricating high-precision sensors in the aerospace industry for component health monitoring. Embedded sensors allow monitoring of the structural integrity for real-time damage detection and assessment. Key developments of the various types of 3D-printed sensors are discussed, highlighting the advantages of this technology. The paper concludes with an assessment of AJP sensor technology in terms of material compatibility, reliability, surface finishing, and manufacturing scale-up.

Keywords: aircraft, aerospace, engines, aerosol jet printing, surface finishing

1. Introduction

The aerospace industry continuously seeks such innovative manufacturing techniques to enhance the performance, reliability, and functionality of its airframe and engine systems. Aerosol jet printing (AJP) has recently gained attention as a non-contact, mask-less, additive manufacturing technique based on the droplet-deposition method. AJP is capable of printing electronic ink with fine pitch and high resolution. Patterning of the functional inks is done by additively depositing at desired location on substrates.

Aerosol jet printing is a unique aerodynamic focusing technique and is capable of printing resolution as small as 10 to 20 μm [1]. AJP can accomplish conformal printing [2] and accommodates a wide range of materials such as graphene [3], carbon nanotube [4], and metals [1–5]. Commonly printed parts are antennas, capacitors, resistors, sensors, and thin-film transistors.

This paper explores the potential of AJP in fabricating sensors for aero-engine components, such as additive manufactured (AM) turbine blades, offering insights into its operational principles, key advantages, and recent advancements.

2. Principles of aerosol jet printing

The Aerosol Jet is a direct-write technology that deposits materials in a precise and controlled manner. The ink is placed into an atomizer, which creates a dense mist of material-loaded droplets between 2 and 5 μm in diameter. The aerosol mist is then delivered to the deposition head, where a sheath gas focuses it. The sheath gas and aerosol accelerate when they pass through the nozzle, forming fine droplets in the flow stream. The sheath gas also prevents the nozzle from clogging [6]. The gas flow and directed towards the substrate through a converging nozzle.

The substrate is positioned below the nozzle on a motion-controlled heated stage, and patterning is achieved by the relative movement of the substrate and deposition head [7–10]. The process is non-contact, enabling features to be printed onto both 2D and 3D surfaces. AJP is therefore capable of patterning functional materials over challenging geometries, which would not be possible with other printing processes. The technology has been reported as a deposition method for functional layers in a diverse range of applications including strain gauges (**Figure 1**) [12].

The process is summarized in the following steps:

1. Atomization: The material ink, typically consisting of nanoparticles or organic molecules, is aerosolized to form a fine mist.

2. Transport: The aerosol mist is transported through a carrier gas to the deposition head.

3. Focusing: The aerosol stream is focused using a sheath gas, allowing for high-resolution patterning on the substrate.

4. Deposition: The focused aerosol stream is deposited onto the substrate, forming the desired patterns.

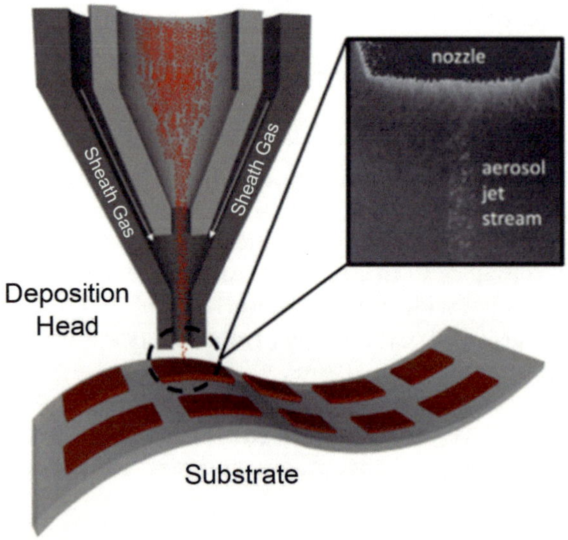

Figure 1.
Illustration of printing an array of flexible electronics on a flexible substrate with the Optomec AJ200 system [11].

The resulting high velocity particle stream remains focused during its travel from the nozzle to the substrate over 2 to 5 mm, maintaining feature resolution on non-uniform substrates. AJP offers several advantages over traditional sensor manufacturing techniques, including the following:

- High-resolution printing (features down to 10 microns).
- Capability to print on non-planar and flexible substrates.
- Low-temperature processing suitable for a wide range of materials.

2.1 The importance of adhesion strength

Maintaining strong adhesion between sensors and structural materials presents several challenges. It is important to ensure that the adhesion strength between the sensor and substrate is sufficiently high to ensure high sensitivity and reliability of the printed sensors. The interfacial bond is influenced by several factors such as substrate cleanliness, surface roughness, printing strategy, and ink quality.

Seifert et al. compared differences between inkjet and Aerosol Jet [13], but none of the results explain the influence of print parameters to gain an understanding of multi-material variability. Agarwala et al. investigated the effect of ultrasonic current and atomizer flow on the print quality of a curved dome shaped structure and established the best parameters for producing homogenous, continuous tracks, free of cracks (**Figure 2**) [6].

Sensing-capable components undergo critical mechanical and electrical changes when stress is applied. Karabal et al. explored various printing strategies with carbon nanotube (CNT)-reinforced polyetherimide (PEI). Two substrates were compared: 3D-orthogonal composite and 3D-orthogonal woven fabric impregnated with epoxy

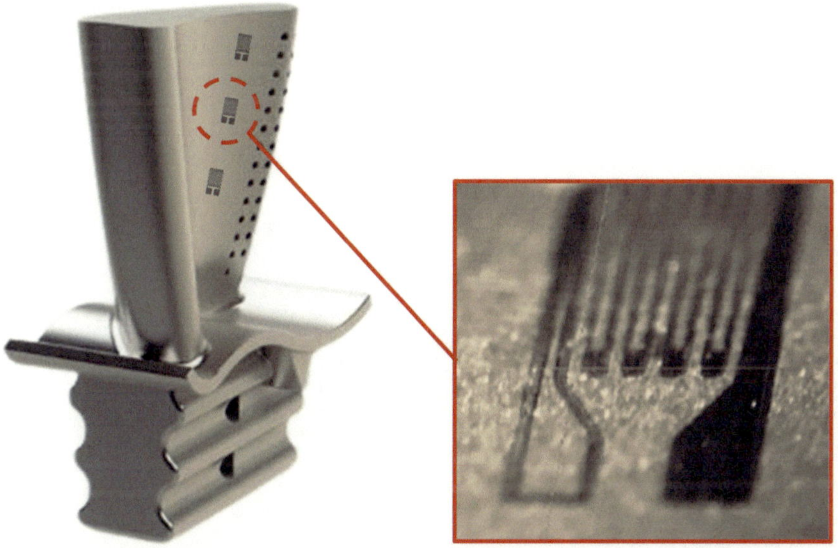

Figure 2.
Illustration of an AJP strain sensor on an additive manufactured turbine blade.

resin. The sensors on the composite did not provide reliable responses due to poor interfacial bonding to the substrate, after cyclic tensile tests. However, the sensors in the woven fabric obtained better piezo-resistive characteristics after cyclic tensile tests [14].

2.2 Surface characteristics of the substrate

Aerosol Jet systems can deposit a wide variety of materials such as polymers, silicon, glass, metals, and ceramics. Femto-liter-sized ink droplets are dispensed by the process which allows for very thin coatings. This leads to good interaction between differently applied layers. Since many electronics materials are expensive, AJP remains an eco-friendly technology which can minimize material use and lower waste disposal costs [15].

Werum et al. studied the importance of material compatibility between ink and substrate by combining Aerosol Jet printing of nano silver inks, photonic sintering, and conductive adhesive bonding. The silver layers on additive manufactured substrates were reinforced by electroless copper plating. This functionalized approach reduced the electrical resistance of the layers. However, the solderability of the part was affected due to crack formation. Detailed exploration is required for better understanding of the relationship between the ink and substrate material [16].

Additive manufactured parts may exhibit some challenging features that complicate the integration AJP sensors. The challenges are mainly as follows:

- High surface roughness and porosity.
- Low thermal stability of the powder materials.
- High substrate coefficient of thermal expansion (CTE) (**Figure 3**).

Typically, Aerosol Jet printing can accommodate rough and porous surfaces. These types of surfaces may have a negative effect on the deposited ink due to relatively high surface energy. This makes it difficult for the printed ink to form a clean and uniform deposit, and the effect is more significant when the surface roughness is larger than the thickness of the ink. A likely solution to this challenge is to pretreat or machine the target area to achieve the desired surface finish before printing.

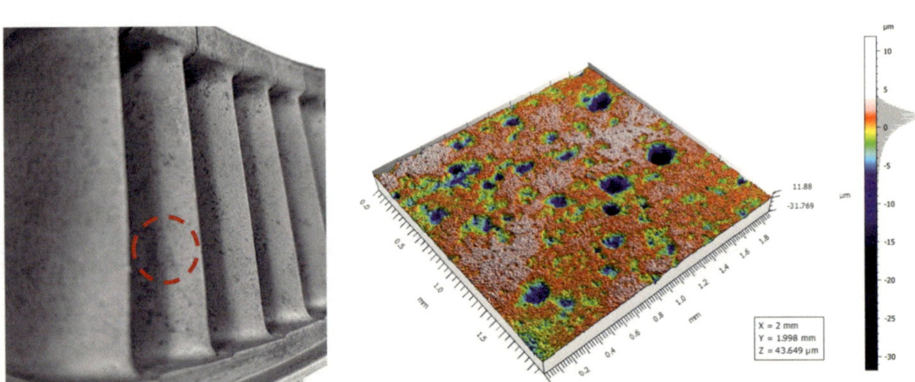

Figure 3.
Example of a compressor vane surface with high porosity.

For substrates with high CTE, thermal mismatch can lead to cracking or delamination of the printed material. Laser sintering may be used to resolve this issue by localized melting of the underlying surface to reduce the residual stresses. Caution should be taken when designing a part to ensure the compatibility of materials properties [17]. An additive manufactured (AM) turbine blade requires printing of high-temperature, super-alloy constituents. Existing AM techniques such as selective laser sintering (SLS) and electron beam melting (EBM) can develop the following surface flaws:

i. Surface artifacts caused by lower process resolution and layering effects.

ii. Granular micro-texture surfaces from melting and adhesion in powder raw materials.

iii. Surface marks due to poor processing or removal of the support structures.

During printing, the layer of the material is often heated and cooled multiple times, resulting in microstructural inhomogeneity, generally known as "fish-scale" morphology. The standard fabricated part has a surface roughness of approximately 3 μm. After post-processing by machining or polishing, the surface roughness can be reduced to approximately 1 μm [18].

2.3 Surface characteristics of the sensor

AJP technology has the capability to generate selective wettability patterns which may be beneficial. When electronic sensors are exposed to sulfur (from the fuel), salt, and other airborne contaminants in the intake air droplets (when flying over seawater), it can lead to engine sulfidation and sensor damage. Traditional manufacturing processes such as etching and anodization tend to be less effective and highly dependent on substrate properties. Coating provides several advantages, such as modifying surface wettability without altering its underlying structural properties. However, these methods are unable to pattern selective wettability surfaces (**Figure 4**) [19].

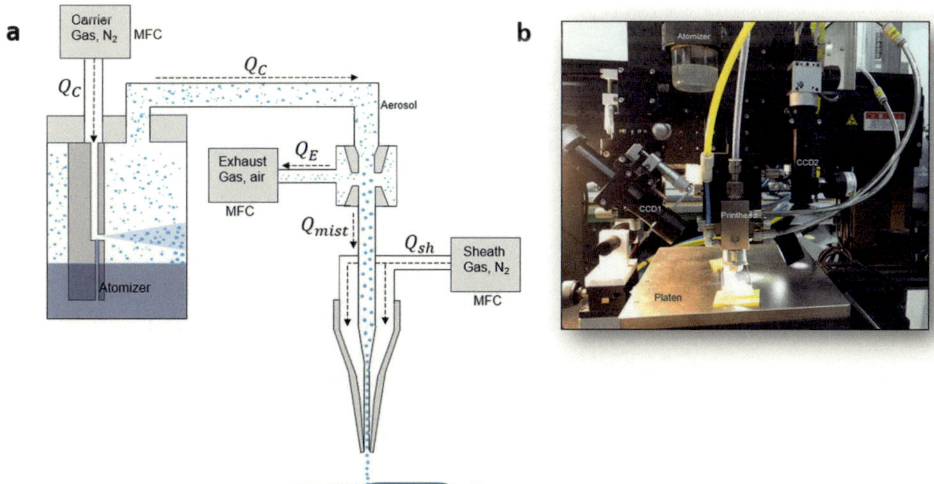

Figure 4.
Experimental framework of the Aerosol Jet. (a) Schematic of the Aerosol Jet printhead. (b) Photograph of the integrated device taken during operation [19].

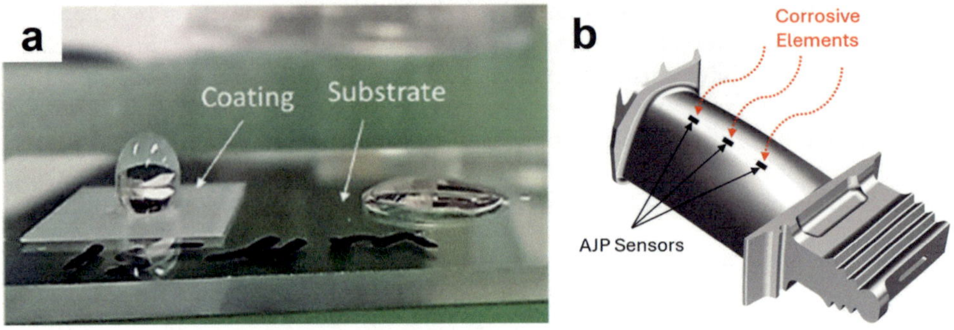

Figure 5.
(a) Wettability of selective area coating. Image of selectively formed wetting/nonwetting pattern on glass substrate using AJP [19] and (b) corrosive elements from the intake air which may damage the engine component and the sensors.

Akuoko et al. developed a method to achieve selective wettability surface patterning using Aerosol Jet printing. This method involved precise deposition of atomized, hydrophobic silica nanoparticle droplets onto substrate surfaces. By directing these atomized droplets, a superhydrophobic surface with a water contact angle of 154° was achieved. This approach holds potential significance in applications requiring waterproofing to safeguard circuit lines against short circuits or damage in corrosive environments from flying over seawater [19].

The printed parallel lines were aligned side by side with predetermined overlays, creating samples with varying line spacings ranging from 60 to 15 μm. This resulted

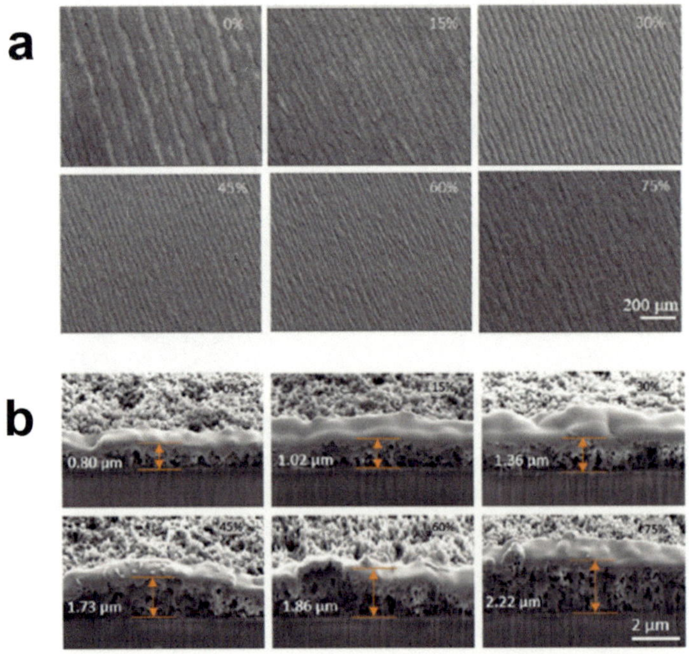

Figure 6.
Micrographs of patterned surface areas. (a) Scanning electron microscope (SEM) images of AJP-coated layers at different percent levels of line overlay. (b) Cross-sectional images of coated layer according to overlapping ratio (from 0 to 75%). These images were obtained through focused ion beam (FIB)-based measurements [19].

in overlays that varied from 0 to 75%. SEM showed that the ridges created an undulating surface structure with a uniform distribution of silica nanoparticles (**Figure 5a**). Further examination of focused ion beam (FIB) images. The average thickness of the grooves (**Figure 5b**) shows an increase from 0.80 to 2.22 μm as the overlapping ratio increased from 0 to 75%, respectively. These observations provide insight into the microstructural characteristics of the printed areas and their relationship with the line overlap (**Figure 6**) [19].

3. Strain gauge sensor applications

AJP is a non-contact, mask-less additive manufacturing method to print electronic ink with fine pitch and high resolution [20]. Strain gauge sensors are essential for detecting deformations and stresses in airfoil components. These devices quantitatively transduce experienced strain or deformations to corresponding changes in resistance, capacitance, or piezoelectricity. Strain gauges have been fabricated using other methods, namely, lithographic printing [21, 22], screen printing [23], flexographic printing [24], and inkjet printing [25] on a wide variety of substrates, but most methods are limited in terms of their printing resolution.

Conventional foil-type strain gauges (**Figure 7**) are made of an active layer, substrate, covering layer, and lead wire. The fabrication process consists of many steps including bonding, curing, UV-exposing, etching, and trimming and is process-intensive. When strain is applied to the active layer, the original geometry of the foil pattern is distorted resulting in a subsequent change in its electrical resistance [27]. Limitations are long measurement range, low sensitivity, and lack flexibility. Thus, it is worth researching new methods to improve the fabrication process.

AJP techniques has many advantages over its counterparts like inkjet and screen printing. AJP can print on 3D surfaces of higher print resolution with 5-axis staging. High-resolution strain gauge with materials such as silver and carbon nanotubes can be effectively printed on various substrates and structures [28–32].

The performance of both conventional and printed strain sensors (**Table 1**), defined as the ratio of relative change in electrical resistance to the mechanical strain, is quantitatively expressed as the gauge factor (GF). The GF of AJP-based resistance-based strain gauges (RSGs) is generally higher than conventional foil-based gauges, with some reports finding that they are capable of an order of magnitude larger GF values [26].

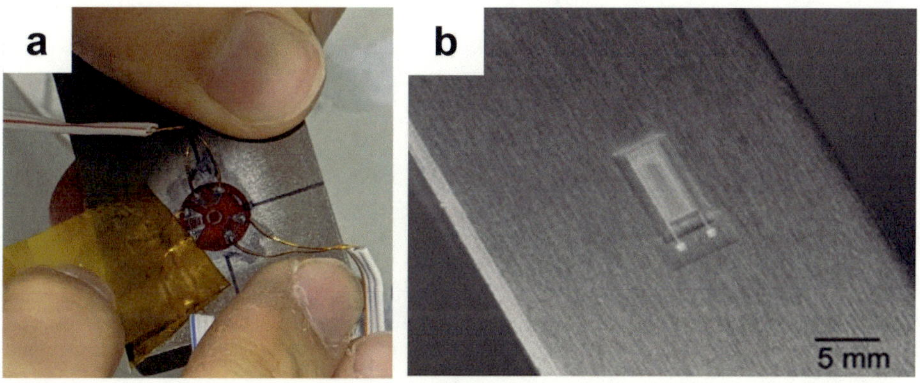

Figure 7.
(a) Conventional foil-type strain gauge and (b) Aerosol Jet-printed strain gauge [26].

Active material	Sensor	Substrate	Sensitivity
(m)-Polyaniline	Strain	Plastic	8.0 GF [a]
Silver	Strain	Aluminum	3.35 GF
Silver	Strain	Carbon fiber	2.2 GF
Silver (NovaCentrix)	Strain	Polyimide	1.74–1.85 GF
Silver (NovaCentrix)	Strain	Polyvinyl chloride	2.42 GF
Silver	Strain	Bandage	N/R [b]
Silver	Strain	Polyimide	N/R
Silver smart ink	Resistance/ capacitance	PET	560 kΩ [c] + 31 pF [d]
Silver	Strain	Carbon composite	N/R
Silver	Strain	Polyimide	5.2 GF
Silver	Strain	Stainless steel	3.15 GF
Silver	Strain	Polyimide	1.65 GF
Silver nanorods + silver nanoparticles	Strain	Polyvinyl alcohol	23 GF
Tellurium + silver nanowires	Piezoelectricity	Silicon	N/R
Silver–palladium	Strain	PZT	N/R
Silver nanowires	Strain	Elastomer	7.5 GF
Silver	Strain	Polyimide	9–10 GF
Silver	Strain	CNT network/ polyimide	20 GF
MWCNT	Strain	Aluminum beam	0.71 GF
PEDOT:PSS	Strain	Polyetherimide beam	0.53 GF
Silver	Strain	Buckypaper (MWCNTs)	1.7 GF
Silver	Strain	Glass	1–5 pF
Silver	Force	Polyimide	4.3 pF N^{-1} [e]
Silver	Strain	PVC + PC + ABS piping	N/R
PDMS/$BaTiO_3$	Pressure	Polyimide	40.52 pF
Carbon	Strain	Elastomer	N/R
Silver	Strain	Aluminum 6061	1.0 GF
Silver	Pressure	Polyimide	1.35 kPa^{-1} [f]
Silver	Strain	Flexible printed circuit	300 Ω–1.2 kΩ
PEDOT:PSS	Pressure	Polyimide	6.2 μV N^{-1} [g]

[a] *GF: gauge factor.*
[b] *N/R: not reported.*
[c] *kΩ: kilo-ohm.*
[d] *pF: picofarad.*
[e] *pF N−1: picofarard/Newton.*
[f] *kPa: 1/kilopascal.*
[g] *μV N−1: microvolt/Newton.*

Table 1.
Collection of ink and substrate types along with best sensitivity values for various AJP-based strain sensors [26].

AJP offers several distinct advantages to the field of microelectronic fabrication (**Figure 8**) [26].

These advantages include the ability to accommodate a wide range of ink viscosities (i.e., 0.001–1 Pa s) [33] and to achieve high-spatial resolution line widths down to approximately 10 μm through fine droplet production, making it a potential alternative to existing approaches (**Table 2**).

Borghetti et al. combined AJP and photonic sintering to produce strain sensors on non-planar surfaces made of temperature-sensitive materials. A protective layer over the printed conductive pattern is often suggested to preserve the sensor and to reduce the influence of external factors of the response signal (**Figure 9**) [35].

Rahman T et al. demonstrated a miniaturized Bluetooth-based wireless, multi-channel readout system compatible with the developed strain gauges for distributed monitoring of soft structural textiles used in space applications [36]. Variations in the ambient temperature can affect the performance of strain sensors by thermal modulations in the electrical properties of printed traces [37].

The strain gauge sensors were tested at high temperature (up to 500°C) to improve accuracy during measurements that could have otherwise been overestimated by

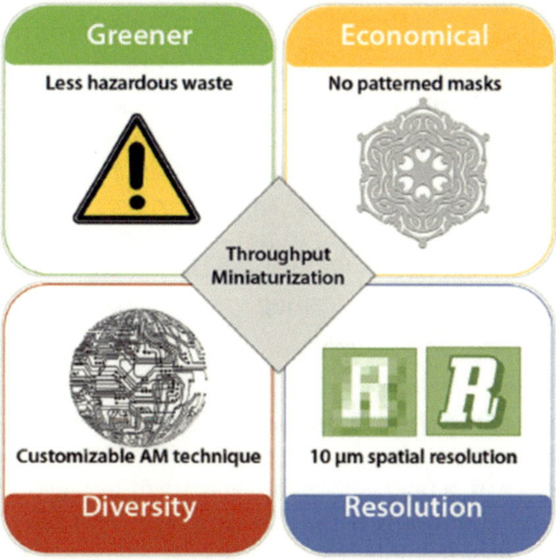

Figure 8.
A summary of advantages conferred by AJP technology [26].

	Dimension	Viscosity [Pa s]	Line width [μm]	Thickness [μm]	Features
Inkjet	2D/3D/4D	0.01–0.02	30–50	0.01–1.0	Uniformity, application variety
Aerosol-jet	2D/3D	0.001–1.0	10–120	0.01–3.0	High resolution, material variety
Screen	2D	0.5–5.0	30–50	0.1–100	Particle load, high viscosity

Table 2.
Comparison of key properties for inkjet, screen, and aerosol-jet printing technologies [34].

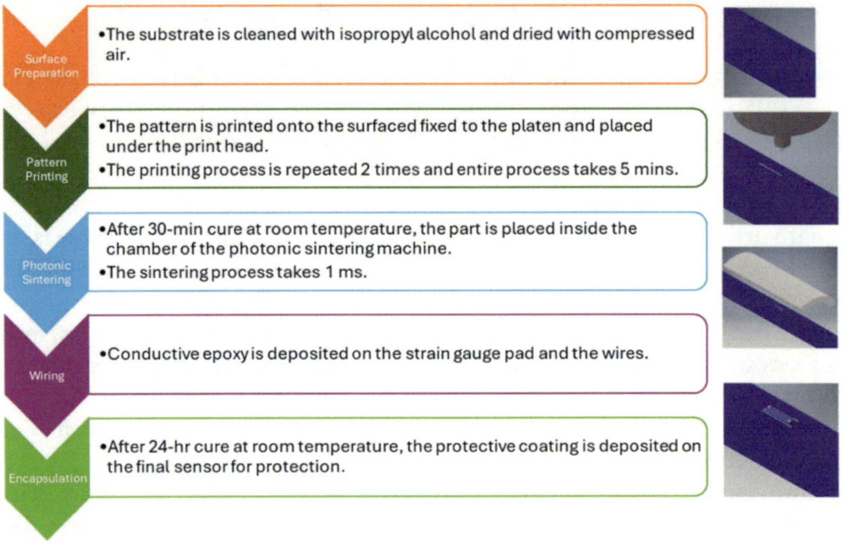

Figure 9.
Manufacturing process for a strain gauge printed directly on a part [35].

temperature-induced changes in the sensing material's resistivity [37, 38]. These temperature-compensated sensors can be used in aerospace applications, where components are exposed to extreme operating conditions.

A common defect in the encapsulation process is the formation of burrs due to masking. Localized burrs may also emerge along the edges of the part upon masking or fixturing removal, where a final layer of paint, erosion polyurethane coating, or thermal barrier coating is applied to the surface. Hence, we shall discuss mitigation strategies to ensure that the surface finishing of the coating is not compromised, and the sensor is not damaged in the process.

4. Surface finishing after the encapsulation process

Upon masking removal, a burr is likely to be formed by the build-up of coating along the edges due to the sidewall of the masking tape. This feature may affect functional performance and compromise the quality of the sensor. Conventional sanding tools are not able to precisely remove burrs without affecting the adjacent areas on the component surface. Micro-blasting is identified as a potential method of removing burrs on the edges of the protective coat (**Figure 10**).

The experiment was conducted to analyze the surface roughness of the coating at the adjacent regions of the micro-blasted samples. The objective is to ensure that the deburred surface will be close to the initial roughness of the substrate.

4.1 Methodology

Two methods for removing burrs on the coat edges by micro-blasting were compared, namely, mask-on and mask-off. Spherical abrasive media, zirconia (ZrO2) of particle size 10–30 μm and hardness of Mohs 7, was used. A blasting nozzle with an inner diameter of 0.46 mm was selected. Process parameters for the experimental setup include a stand-off height of 5 mm and linear speed of 1 mm/s.

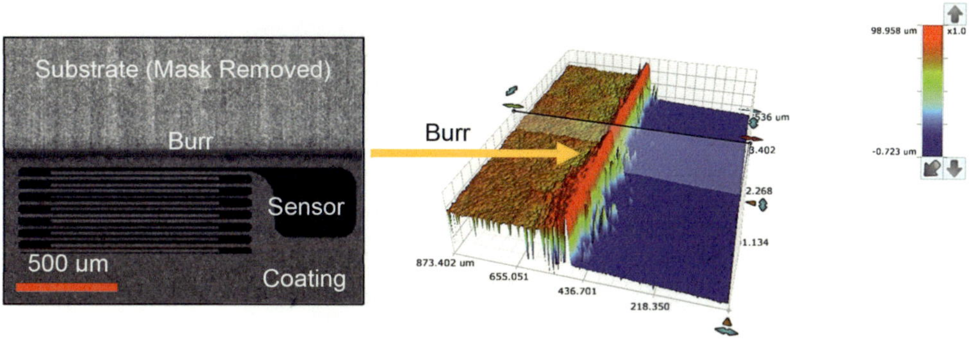

Figure 10.
Surface 3D topography of a coating burr after masking tape removal.

Optical surface measurement is performed by coherence correlation interferometry (CCI), 3x replications along the edge (**Figure 11**). *Sa* is the arithmetic mean of the absolute value of the height within a sampling area and is expressed as follows:

$$Sa = \frac{1}{A}\iint_A |Z(x,y)|\,dxdy \tag{1}$$

4.2 Results

The roughness parameter, *Sa*, shows that the mask-on method (**Figure 12**) produced values close to or lower than the initial condition. The mask-off method produced slightly rougher surfaces (not ideal).

4.3 Discussion and analysis

Although the mask-on method produced ideal Sa values, evaluation of roughness by Sa value alone may not be adequate due to the high variation caused by (i) inconsistencies in the coating thickness and (ii) maskant removal by manual means. Alternatively, fast Fourier transformation (FFT) and power spectral density (PSD) approach could be used as a more detailed characterization.

PSD provides a more reliable description of the topography than the roughness and imparts several useful information of the surface. The spikes in the curve indicate

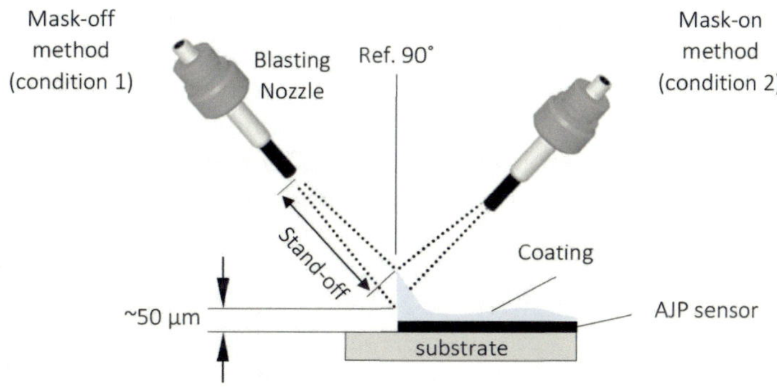

Figure 11.
Micro-blasting strategy for the two methods.

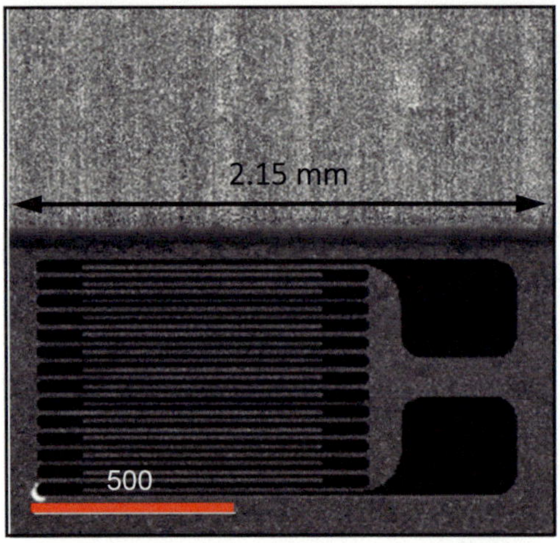

Figure 12.
Location of measurement on the coating side (after micro-blasting) by coherence correlation interferometry (CCI). Square grid dimensions measuring 2.15 mm by 2.15 mm, 3x replications.

that there are high-frequency harmonic components due to the random array of media indentation by micro-blasting.

PSD is a measure of the amplitude of each harmonic part for a specific frequency and along a given direction. Mathematically, PSD is defined as the square of the Fourier transform of the measured surface roughness and is expressed as follows:

$$PSD(f) = \frac{d_0}{N}\left|\sum_{j=1}^{N} Z_j . \exp\left[-i.2\pi f(j-1)d_0\right]\right|^2 \quad (2)$$

where $i = \sqrt{-1}$; d_0 is the sampling length; Z_j is the surface amplitude function, the spatial frequency f equal to K / L, where K is an integer that ranges from 1 to N /2; N is the number of sampling points over a finite length L of surface errors.

Based on Fourier analysis, surface roughness is assumed to be a series of sine waves with different frequencies and amplitudes. This makes it possible to detect smaller spectral structures and to improve roughness direction results. Thus, PSD provides a more reliable description to the topography and imparts several useful information of the surface. The PSD spectrum emerges as color-scaled function values, and the point indicates the magnitude displayed on the z-axis.

The PSD profile shows the plot of power spectral density as a function of spatial frequency (f). The averaged PSD study inverts the horizontal axis ($\lambda = 1/f$) and proposes a linear reading of wavelengths instead of frequencies. dBc (decibels relative to the carrier) is the power ratio of a signal to a carrier signal, expressed in decibels.

From **Figure 13**, the mask-on method produced amplitudes close to or lower than the initial condition for substrate and coating. Mask-off produced a higher amplitude. The higher amplitude is likely due to the nozzle direction (facing the coating) (**Figure 14**).

A general process capability test was performed to compare the two methods. The sampling is based on a population size of 25 (no. of runs) in each group. Cpk (process capability) indicates whether a process can meet specifications. Ppk (process performance) indicates how a process is performed over time. Results are as follows.

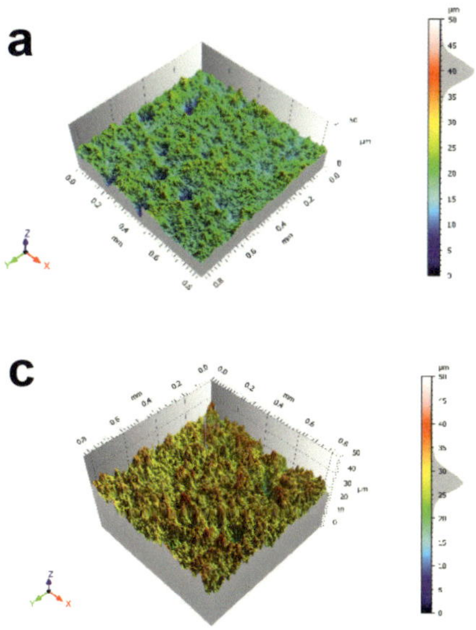

Figure 13.
Surface topography of the coating after micro-blasting. (a) Initial roughness, Sa *of 1.75 μm, (b) mask-on (60° nozzle angle)* Sa *of 1.76 μm, and (c) mask-off (60° nozzle angle)* Sa *of 2.01 μm.*

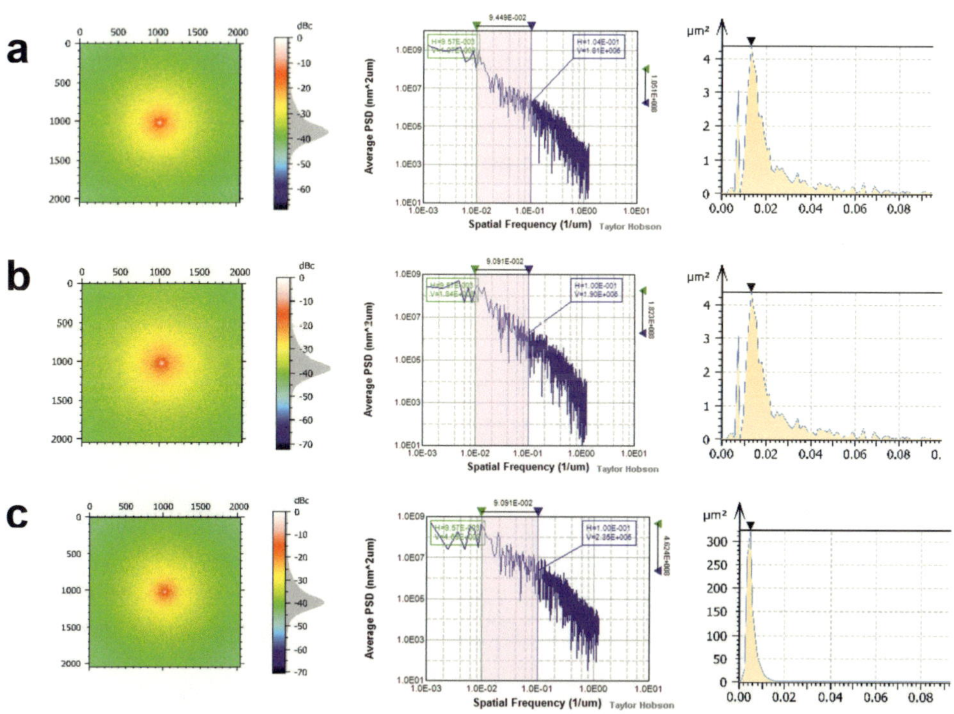

Figure 14.
Power spectral density (PSD) results of the coating surfaces. (a) Initial condition: magnitude: −47.3 dBc, PSD profile and amplitude of 2.09 μm, (b) mask-on (60° nozzle angle): PSD profile and amplitude of 2.09 μm, (c) mask-off (60° nozzle angle): PSD profile and amplitude of 17.95 μm.

4.3.1 Capability histogram

The peak of the distribution curve for Mask-Off method is within the upper specification limit (USL); while the mask-on method is more centered on the target but slightly above the USL. This is due to the inclusion of tape and coating thickness in step height.

4.3.2 Normal probability plot

Hypotheses test by Anderson-Darling (95% confidence interval) - H0: when p-value ≥ 0.05, the data is a normal distribution; H1: when p-value < 0.05, the data is not a normal distribution. Mask-on method produced a p-value of 0.076; therefore, we concluded the data is a normal distribution. Mask-off method indicates more departures from normality and may not be accurate since p-value is < 0.005 (data is non-normal). This is due to the randomness of step height exhibited by manual tape peeling. Non-normal distribution requires "Transform" option to analyze the data.

4.3.3 Process capability

By industry standards, Ppk and Cpk >1.33 indicates a stable process. The mask-off method produces higher Cpk and Ppk values than the mask-on method. A possible reason is that a partial amount of coating is removed in the tape peeling process, thus inadvertently lowering the step height even before blasting is commenced. Both methods require process improvement since Ppk and Cpk are <1.33. Patterns in the data indicate that the variation is due to special causes that should be corrected (**Figures 15** and **16**).

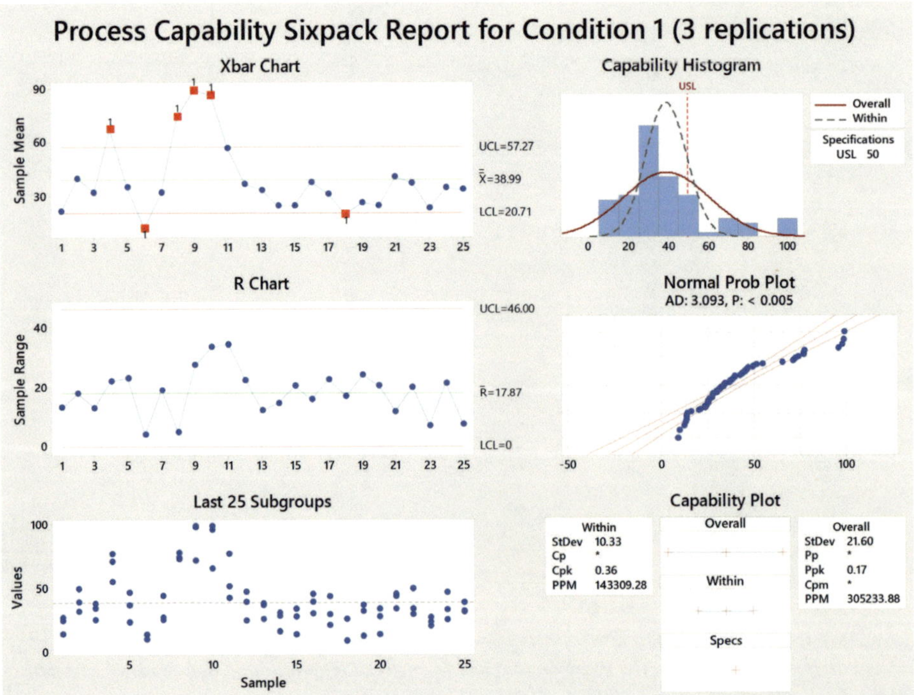

Figure 15.
Process capability (Cpk) for mask-off method (condition 1).

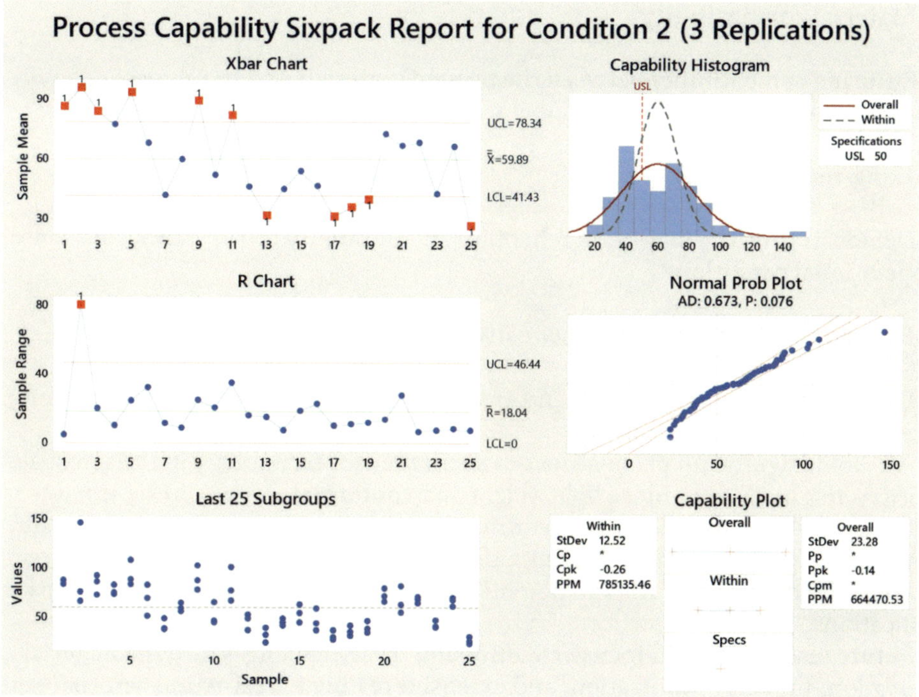

Figure 16.
Process capability (Cpk) for mask-on method (condition 2).

5. Conclusion

Aerosol Jet printing (AJP) has emerged as a powerful technique to print flexible sensors on various substrates for measuring quantities like stress, torque, strain, pressure, and vibration. Nano-sized strain gauges have higher sensitivity and better precision, thus leading to enhanced performance. As technology matures, the sensors need to be smarter, efficient, lighter, and robust.

AJP offers several distinct advantages for aerospace sensor fabrication.

Lightweight and Low-Profile Designs:

The ability to print thin, conformal layers reduces the weight and space requirements of sensor systems.

Integration and Miniaturization:

The integration of multiple sensor types onto a single substrate, leading to compact and multifunctional sensor networks.

Customization and Rapid Prototyping:

Easy customization and rapid prototyping of sensor designs, accelerating development cycles.

6. Challenges and future directions

Despite its advantages, AJP faces several challenges that need to be addressed to fully realize its potential in aerospace applications.

6.1 Material compatibility

Ensuring compatibility between the printed materials and the aerospace environment, including factors such as temperature extremes and mechanical stresses.

6.2 Long-term reliability

Demonstrating the long-term reliability and durability of AJP-fabricated sensors in operational conditions.

6.3 Scale-up and manufacturing consistency

Developing scalable processes and ensuring consistency in sensor performance across large production volumes.

AJP holds significant promise for advancing sensor technology in the aerospace industry. Its ability to produce lightweight and conformal sensors presents new opportunities for enhancing the performance and reliability of aerospace systems. Continued research and development efforts will be crucial in overcoming current challenges and unlocking the full potential of AJP-fabricated sensors for aerospace applications.

Future research should focus on addressing these challenges through material innovation, process optimization, and extensive testing in real-world aerospace environments. Additionally, exploring hybrid manufacturing approaches that combine AJP with other additive and subtractive techniques could enhance the capabilities and applications of printed sensors.

Acknowledgements

A research paper on Aerosol Jet Printing, School of Mechanical & Aerospace Engineering, Nanyang Technological University.

Conflict of interest

The author declares no conflict of interest.

Author details

Edgar Jeevan Danaraj
Nanyang Technological University, Singapore

*Address all correspondence to: edgarjee001@e.ntu.edu.sg

References

[1] Wang K, Chang Y-H, Zhang C, Wang B. Evaluation of quality of printed strain sensors for composite structural health monitoring applications. International SAMPE Technical Conference. 2013. Available from: https://www.researchgate.net/publication/288108506

[2] Paulsen JA, Renn M, Christenson K, Plourde R. Printing conformal electronics on 3D structures with aerosol jet technology. In: Future of Instrumentation International Workshop (FIIW), 2012. IEEE; 2012

[3] Song D, Mahajan A, Secor EB, Hersam MC, Francis LF, Frisbie CD. High-resolution transfer printing of graphene lines for fully printed, flexible electronics. ACS Nano. 2017;**11**(7):7431-7439

[4] Cao C, Andrews JB, Franklin AD. Completely printed, flexible, stable, and hysteresis free carbon nanotube thin film transistors via aerosol jet printing. Advanced Electronic Materials. 2017;**3**(5):1700057

[5] Krzeminski J, Wroblewski G, Dybowska-Sarapuk L, Lepak S, Jakubowska M. Nanosilver conductive lines made by spray coating and aerosol jet printing technique. In: Photonics Applications in Astronomy, Communications, Industry, and High Energy Physics Experiments 2017. International Society for Optics and Photonics; 2017

[6] Agarwala S, Goh GL, Yee YW. Optimizing aerosol jet printing process of silver ink for printed electronics. IOP Conference Series: Materials Science and Engineering. 2017;**191**:012027

[7] Mahajan A, Frisbie CD, Francis LF. Optimization of aerosol jet printing for high-resolution, high-aspect ratio silver lines. ACS Applied Materials & Interfaces. 2013;**5**(11):4856-4864

[8] Smith M, Choi YS, Boughey C, Kar-Narayan S. Controlling and assessing the quality of aerosol jet printed features for large area and flexible electronics. Flexible and Printed Electronics. 2017;**2**(1):27

[9] Binder S, Glatthaar M, Rädlein E. Analytical investigation of aerosol jetprinting. Aerosol Science and Technology. 2014;**48**(9):924-929

[10] Mashayekhi M, Winchester L, Evans L, Pease T, Laurila M-M, Mäntysalo M, et al. Evaluation of aerosol, superfine inkjet, and photolithography printing techniques for metallization of application specific printed electronic circuits. IEEE Transactions on Electron Devices. 2016;**63**(3):1246-1253

[11] University of Wisconsin–Madison, Andrews Laboratory for Printed Electronics and Sensors. Section: Printed Electronics, point number 3. Available from: https://andrews.engr.wisc.edu/research/

[12] Maiwald M, Werner C, Busse M, Zoellmer V. INKtelligent printed strain gauges. Sensors and Actuators A: Physical. 2010;**162, 2**:198-201

[13] Seifert T, Sowade E, Roscher F, Wiemer M, Gessner T, Baumann RR. Industrial & Engineering Chemistry Research. 2015;**54**:769

[14] Karabal M, Dincer M, Arslan D, Yüksel R, Akyazici ZN, Yildiz A, et al. Direct 3D printing of strain sensors

onto 3D woven orthogonal composite structures: Evaluating two distinct approaches for sensor performance. In: AIAA SCITECH 2024 Forum. (n.d.). DOI: 10.2514/6.2024-2284

[15] King B, Renn M. Aerosol Jet® Direct Write Printing for mil-Aero Electronic Applications. USA: Optomec, Inc.; 2019

[16] Werum K, Mueller E, Keck J, Jaeger J, Horter T, Glaeser K, et al. Aerosol jet printing and interconnection technologies on additive manufactured substrates. Journal of Manufacturing and Materials Processing. 2022;**6**(5):119. DOI: 10.3390/jmmp6050119

[17] Hedges M, Borras A. Marin 3D Aerosol Jet. Printing - adding electronics functionality to RP/RM, DDMC, page 4. 2012 Conference, 14-15.3.12, Berlin. Available from: https://www.optomec.com/wp-content/uploads/2014/04/Optomec_NEOTECH_DDMC_3D_Aerosol_Jet_Printing.pdf

[18] Sinha A, Swain B, Behera A, Mallick P, Samal SK, Vishwanatha HM, et al. A review on the processing of aero-turbine blade using 3D print techniques. Journal of Manufacturing and Materials Processing. 2022;**6**(1):16. DOI: 10.3390/jmmp6010016

[19] Akuoko SY, Mosa MA, Jo JY, Lee J, Kwon K-S. Achieving selective wettability surface through aerosol jet hydrophobic line printing. Figure 5(a), 5(b), 6(a), 10(a), 10(b). ACS Omega. 2024;**9**(5):5661-5674. DOI: 10.1021/acsomega.3c08003

[20] Tan H, Tran T, Chua C. A review of printed passive electronic components through fully additive manufacturing methods. Virtual and Physical Prototyping. 2016;**11**(4):271-288

[21] Hay GI, Evans PSA, Harrison DJ, Southee D, Simpson G, Harrey PM. Characterization of lithographically printed resistive strain gauges. IEEE Sensors Journal. 2005;**5**(5):864871

[22] Hay GI, Southee DJ, Evans PS, Harrison DJ, Simpson G, Ramsey BJ. Examination of silver-graphite lithographically printed resistive strain sensors. Sensors and Actuators A: Physical. 2007;**135**(2):534-546

[23] Bessonov A, Kirikova M, Haque S, Gartseev I, Bailey MJA. Highly reproducible printable graphite strain gauges for flexible devices. Sensors and Actuators A: Physical. 2014;**206**:75-80

[24] Maddipatla D, Narakathu B, Avuthu SGR, Emamian S, Eshkeiti A, Chlaihawi A, et al. A novel flexographic printed strain gauge on paper platform. 2015:1-4. DOI: 10.1109/ICSENS.2015.7370606

[25] Andò B, Baglio S. All-inkjet printed strain sensors. IEEE Sensors Journal. 2013;**13**(12):4874-4879

[26] Fisher C, Skolrood LN, Li K, Joshi PC, Aytug T. Aerosol-jet printed sensors for environmental, safety, and health monitoring: A review. Advanced Materials Technologies. 2023;**8**:2300030. Figure 1. Table 3. DOI: 10.1002/admt.202300030

[27] Maiwald M, Werner C, Zöllmer V, Busse M. Sensor Review. 2010;**30**:19

[28] Li S, Park JG, Wang S, Liang R, Zhang C, Wang B. Working mechanisms of strain sensors utilizing aligned carbon nanotube network and aerosol jet printed electrodes. Carbon. Jul. 2014;**73**:303-309

[29] Blumenthal T, Fratello V, Nino G, Ritala K. Conformal printing of sensors on 3D and flexible surfaces using aerosol jet deposition. Proceedings of SPIE. 2013;**8691**:86910P

[30] Zhao D, Liu T, Zhang M, Liang R, Wang B. Fabrication and characterization of aerosol-jet printed strain sensors for multifunctional composite structures. Smart Materials and Structures. 2012;**21**(11):115008

[31] Thompson B, Yoon H-S. Aerosol printed carbon nanotube strain sensor. Proceedings of SPIE. 2012;**8346**:83461C

[32] Thompson B, Yoon H-S. Aerosol-printed strain sensor using PEDOT: PSS. IEEE Sensors Journal. 2013;**13**(11):4256-4263

[33] Wilkinson NJ, Smith MAA, Kay RW, Harris RA. International Journal of Advanced Manufacturing Technology. 2019;**105**:4599

[34] Fisher C, Skolrood LN, Li K, Joshi PC, Aytug T. Aerosol-jet printed sensors for environmental, safety, and health monitoring: A review. Advanced Materials Technologies. 2023;**8**:2300030. Table 2. DOI: 10.1002/admt.202300030

[35] Borghetti M, Serpelloni M, Sardini E. Printed strain gauge on 3D and low melting point plastic surface by aerosol jet printing and photonic curing. Sensors. Figure 2. 2019;**19**(19):4220. DOI: 10.3390/s19194220

[36] Ranganatha KL, Novich K, Phero T, Fujimoto KT, Litteken D, Estrada D, et al. IEEE Sensors. 2021;**2021**:9639749

[37] Rahman T, Moser R, Zbib HM, Ramana CV, Panat R. 3D printed high performance strain sensors for high temperature applications. Journal of Applied Physics. 2018;**123**:024501

[38] Phero TL, Novich KA, Johnson BC, McMurtrey MD, Estrada D, Jaques BJ. Sensors and Actuators, A: Physical. 2022;**344**:113691

Chapter 7

Challenges in Electrical Insulation Materials and Thermal Management for Medium Voltage Power Cables for Envisaged Wide-Body All-Electric Aircraft

Anoy Saha and Mona Ghassemi

Abstract

All-electric aircraft (AEA) emergence is considered a promising initiative toward achieving net-zero aviation. Future wide-body AEA will require electric power systems (EPS) with high power density and minimal system mass. Power cables, a crucial element of the aircraft EPS, need to be designed to enhance the EPS's overall power density. At the cruising altitudes of wide-body AEA, the limited heat transfer by convection poses significant thermal challenges for the design of power cables. The challenges are further intensified when employing bipolar medium voltage direct current (MVDC) EPSs, typically consisting of two power cables, negative and positive poles, positioned adjacent. The cable's surface area influences both radiative and convective heat transfers. This book chapter deals with the design, fabrication, and testing of aircraft MVDC power cables. Multilayer multifunctional electrical insulation (MMEI) systems were recently introduced instead of single-layer insulation in the aforementioned cables, which are discussed. In addition to delineating coupled electrical, thermal, and computational fluid dynamic models to obtain thermal distribution and electric stress within the cable and using the model for optimal design of cable and duct geometries, all modeling details in COMSOL Multiphysics are also explained, resulting in this chapter book as a textbook and valuable reference.

Keywords: wide-body all-electric aircraft, power cables, medium voltage, electrical insulation, thermal management, low pressure, convective heat transfer

1. Introduction

Moving towards net-zero emissions in transportation necessitates a comprehensive shift to all-electric solutions, including in aviation, where innovations are crucial. The

UN's Sustainable Development Goals aim for urgent action on climate change, targeting a 43% reduction in GHG emissions by 2030 and net zero by 2050 [1]. With transportation accounting for 29% of U.S. GHG emissions in 2021, including 8% from aviation, electrification is the most practical route to achieve net-zero emissions [2, 3]. The rapid advancement in battery technology and renewable energy integration, along with strong commitments from companies and governments, is greatly accelerating the adoption of electric vehicles (EVs), thereby reducing greenhouse gas emissions and fossil fuel dependency [4, 5]. The journey to net zero is more complex but equally crucial in aviation. The aviation sector, with a historical annual growth rate of 4–5%, presents a unique challenge, with over 80% of its greenhouse gas emissions originating from narrow and wide-body aircraft [6, 7]. Emerging technologies, such as electric and hybrid-electric-aircraft, promise to decrease emissions, with several prototypes already demonstrating viable short-haul flights [8]. The International Air Transport Association (IATA) has set ambitious targets for net-zero aviation by 2050, emphasizing the need for sustainable aviation fuels (SAFs) alongside electrification to achieve these goals [9].

Although there have been considerable advancements in electric vehicles and other electrified transportation, the process of electrifying commercial wide-body airplanes is still in its initial phases. There are two main types of electrified aircraft: more electric aircraft (MEA) and all-electric aircraft (AEA). MEA uses conventional combustion propulsion systems but replaces pneumatic, hydraulic, and mechanical systems with electrical counterparts to enhance overall efficiency [10]. AEA, on the other hand, includes all the electrified systems of MEA and features an electrified propulsion system. Conventional aircraft's electrified propulsion systems require a power density of 40 kVA/kg. Achieving this power density necessitates using EPS with high power density and low system mass [11, 12].

An aircraft's EPS includes components like power electronic devices, electric machines, electrochemical energy units (EEUs), and multiple loads, all connected by power cables. These cables contribute significantly to the system's total mass, accounting for more than 30% [10]. Optimizing the cable systems is crucial to creating EPS with high power density and low system mass. Operating at higher voltages can effectively reduce cable weight, but this introduces substantial challenges in designing appropriate cable insulation. The design of power cable systems requires accurate, current carrying capacity calculations, which can be challenging under complex environmental and geometric conditions. While analytical approaches per IEC 60287 are effective in homogeneous and simple scenarios, they struggle with heterogeneous environments, varied material thermal properties, and non-constant temperature limits. In such cases, numerical methods like the finite element method (FEM) are essential. FEM, as highlighted by standards such as IEC TR 62095, enables engineers to conduct detailed analyses of both electromagnetic and thermal behaviors within power cables. It enables precise temperature distribution and heat dissipation analysis, crucial for determining current ratings in different conditions [13–15].

This chapter of the book will explore the challenges related to insulation systems in aircraft cable systems. Section 2 delves into a detailed exploration of MEA and AEA EPS architecture. Section 3 covers the modeling and simulation of bipolar cable systems. Section 4 elaborates on the cable design modeling process using COMSOL Multiphysics software. Finally, Section 5 outlines the procedures for fabricating and testing cable insulation systems.

2. Overview of MEA and AEA electric power system (EPS) architecture

2.1 MEA EPS architecture (past to present)

The MEA concept in aircraft design focuses on substituting traditional pneumatic and hydraulic systems with electrical systems. **Figure 1** shows the progression of electrical power capacity in commercial aircraft from the early 20th century to the present day. Starting with minimal power in the Wright Flyer, there's a steady increase with models like the DC-3 and B707, followed by a significant escalation with the B737 in 1966. This growth continues with models like the B767, A320, B747–400, and A340, reaching 200–400 kVA capacities. The trend peaks with modern planes like the A380 and Boeing 787, which achieve impressive capacities, notably the 787 reaching 1000 kVA by 2010 [12]. These advancements reflect the industry's shift towards MEA for improved efficiency, reduced weight, and better performance.

A typical commercial airplane utilizes a 115-V AC voltage at 400 Hz, supplied by a generator connected to the main engine, to maintain a constant frequency. Air circulation fans, avionics, lights, and galley equipment are the main uses of electric power in that constant voltage and constant frequency architecture [16–18]. The conventional aircraft also has a 28-V DC bus, derived from the 115-V AC via transformer rectifier units (TRUs), which is further reduced within line replaceable units (LRUs) for lower voltage requirements such as 5 V and 3.3 V, used for integrated circuits and microprocessors [19].

Aircraft electrical systems employ multiple busses for redundancy, interconnected by tiebreakers, with switches managing generators, loads, and busses. The modern trend in aircraft, like the Boeing 787 and Airbus A380, replaces traditional constant voltage constant frequency (CVCF) systems with constant voltage variable frequency

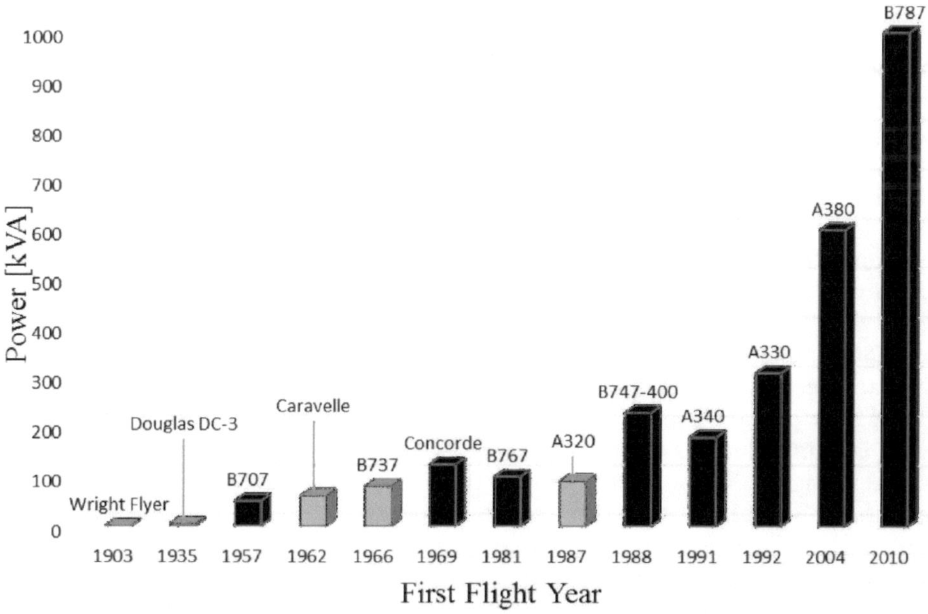

Figure 1.
Evolution of electrical power requirements: Gray represents short- to medium-range aircraft, and black represents medium- to long-range aircraft [12].

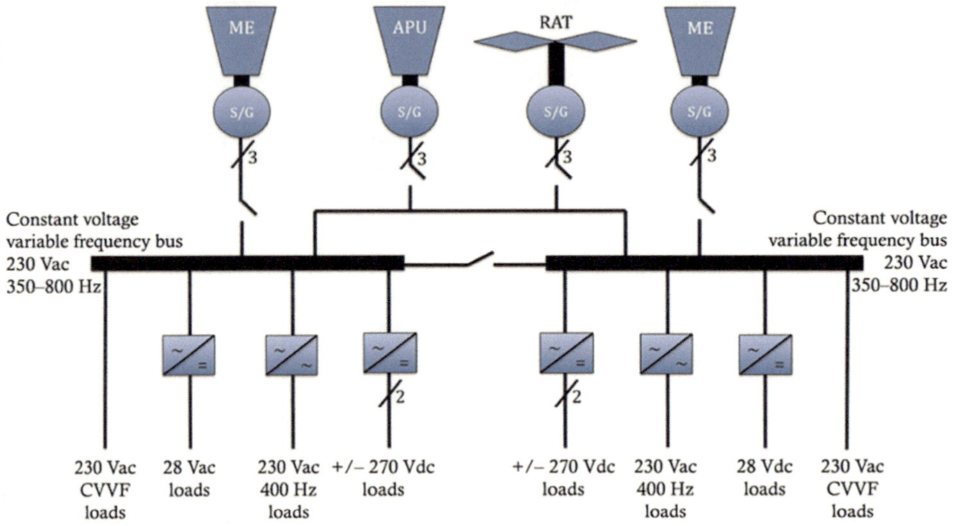

Figure 2.
CVVF bus power system utilized in MEA [19].

(CVVF) systems [19]. These maintain regulated voltage (115 or 230 VAC) while varying frequency (350–800 Hz) based on engine speed. **Figure 2** illustrates the CVVF bus power system utilized in MEA.

The Airbus A380 integrates advanced "more electric" technology, including four variable-frequency electrical generators delivering 150 kVA each, enhancing power reliability and efficiency. This system supports the aircraft's extensive electrical needs with flexibility in power output. The aircraft's hydraulic systems have been upgraded with electrical systems 1 and 2, replacing traditional hydraulics with electrical backup hydraulic actuators (EBHA) and electro-hydrostatic actuators (EHA) [20–22]. These actuators combine electrical and hydraulic power for lighter weight and increased redundancy in flight controls, ensuring operational reliability even under hydraulic system failures. Airbus's adoption of EHAs marks a significant advancement in aviation technology, demonstrating a move towards more efficient and reliable aircraft systems.

The Boeing 787 Dreamliner employs a four-channel variable frequency starter generator (VFSG) system featuring two 250 kVA VFSGs on each main engine. This innovative electrical system replaces traditional pneumatic systems, delivering enhanced efficiency and weight savings by eliminating heavy bleed air components. The VFSG simplifies auxiliary power unit (APU) design, operating as a shaft power-only machine and improving compatibility with ground support infrastructure. The VFSG is a six-pole, brushless, three-phase, alternating current synchronous machine housed in aluminum and driven directly from the main engine gearbox. It has a nominal rating of 235 volts alternating current (VAC), 250 kVA, three phases, and an output frequency of 360–800 Hz [23–25]. This design ensures more efficient power distribution and use, with approximately 20 miles (32 km) less wiring than the 767. Additionally, the 787's electrical system allows for autonomous engine start without external power, and backup sources, including main and APU batteries, ensure continued operation in case of power failure. These advancements ensure better fuel efficiency, lower maintenance costs, and compliance with safety regulations, positioning the 787 as a cutting-edge aircraft in modern aviation.

2.2 Concept design of AEA

Researchers have developed a range of innovative aircraft concepts intending to make electric flight a reality for commercial transport. To support this initiative, NASA and its industry collaborators have undertaken design studies to assess the potential decrease in fuel consumption and emissions and investigate the technological prerequisites for different configurations across a range of sizes, markets, and time periods.

NASA's SUbsonic Single Aft eNgine (SUSAN) Electrofan is an advanced hybrid-electric aircraft designed to cut emissions by 50% within the next few decades, aiming to net-zero emissions. It can carry 180 passengers with a 750-mile economic range, utilizing a 20 MW Electrified Aircraft Propulsion (EAP) system and advanced propulsion airframe integration (PAI) for improved aerodynamic efficiency. SUSAN incorporates technologies like the 1.4 MW High-Efficiency Megawatt Motor (HEMM) and the High-Efficiency Electrified Aircraft Thermal Research (HEATheR) power converter, aiming to revolutionize sustainable aviation while operating within existing airport infrastructures [26].

The STARC-ABL, developed by NASA, is an innovative electric aircraft concept that utilizes Boundary Layer Ingestion (BLI) to enhance aerodynamics and minimize drag. The turboelectric propulsion system of this aircraft incorporates under-wing turbofan engines that serve as generators for a rear motor. This design is intended to reduce fuel consumption. Featuring a 2–3 MW power system with the HEMM, STARC-ABL aims to showcase substantial advancements in electric propulsion for cleaner and more efficient air travel solutions [27].

The NASA N3-X features a hybrid wing body design with turboelectric distributed propulsion (TeDP), employing lightweight superconducting electric motors and generators. Sixteen distributed fans manage boundary layer airflow for improved aerodynamic efficiency and reduced noise, significantly lowering fuel consumption and emissions. This technology enables quieter operations and facilitates the use of smaller airports, marking a significant step towards sustainable and efficient commercial aviation [28].

Empirical Systems Aerospace's (ESAero) ECO-150 concept employs a turboelectric setup with distributed electric propulsion, utilizing two large turboshaft generators to power electric ducted fans integrated into the wing surfaces. This split-wing configuration enhances aerodynamic efficiency and reduces noise through embedded propulsors. The design targets improved fuel efficiency, emissions reduction, and noise mitigation in alignment with NASA's N + 2 goals for advanced aircraft technologies [29].

3. Model and design of bipolar cable systems

As a crucial component of the aircraft EPS, power cables offer significant potential for reducing the overall system mass through improvements. Medium voltage (MV) power cables operating within the range of several kilovolts meet high power density requirements. However, increasing voltage necessitates thicker insulation, resulting in significant weight gain. Designing power cables for aircraft applications presents various challenges, including surface charge management, arc and arc tracking prevention, partial discharge (PD) mitigation, and efficient thermal management [12]. Thus, lightweight MV cables resistant to PD and arcs are crucial for electrified propulsion systems. In bipolar cable configuration, these challenges are much more critical. Achieving lightweight MV cables suitable for aircraft systems involves two approaches: redesigning the cable insulation system to enhance performance and reliability under aviation's challenging conditions and modifying the cable's shape to improve heat dissipation capabilities. These efforts aim to create lightweight MVDC power cables that meet the critical requirements of aircraft applications.

3.1 Challenges associated with designing aircraft cable systems

Arc tracking is a critical failure mode in electrical cable systems, characterized by forming a conductive path along the surface of insulating materials, leading to carbonization and eventual electrical breakdown. This phenomenon is particularly concerning in high-voltage environments, where the risks of arcing are elevated due to higher voltage levels, increased dv/dt (rate of voltage change), higher power densities, and reduced wire distances. The chemical composition of the insulation is a key factor in determining its susceptibility to tracking, with PDs being the main cause [30]. Arc tracking in aircraft cables can be classified into two main categories based on the environmental conditions and mechanisms involved: wet arc tracking and dry arc tracking. Wet arc tracking occurs when moisture or fluids create a conductive path between exposed wires or to the aircraft structure, leading to short circuits and carbonization. This is common in humid or fluid-exposed environments. Dry arc tracking results from insulation degradation due to aging, UV radiation, abrasion, and poor installation, causing localized heating and erosion [31].

PDs are localized electrical discharges that partially bridge the insulation between conductors. They pose a critical concern in aircraft electrical systems, especially during rapid ascents and descents when lower air pressures and increased moisture condensation exacerbate their occurrence. Lower air pressure enhances both the magnitude and frequency of PDs while reducing the partial discharge inception voltage (PDIV), particularly at higher operating voltages. Furthermore, higher voltage frequencies further decrease PDIV, increasing the likelihood of PDs [32, 33]. To manage PDs effectively, aircraft systems can utilize screened cables or incorporate corona-barrier materials in unscreened cables [34]. Corona-barrier materials significantly enhance the system's resistance to PDs, prolonging the time to failure compared to conventional insulation materials. While they do not eliminate PD-induced failures, they notably delay the onset of failure, thereby improving system reliability. Addressing PD challenges in aircraft electrical systems involves developing insulation systems with screened cables or integrating corona-barrier materials, which effectively mitigate PD effects and enhance the durability and performance of cable systems.

Surface discharges in polymer insulation occur when the electric field at the insulator surface becomes strong enough to ionize surrounding air, resulting in PDs. This typically arises from air gaps or dry bands on the insulator surface, often due to hydrophobicity loss, contaminants, or moisture presence. When the voltage gradient surpasses a critical level, it triggers surface discharges, generating localized heating that can carbonize the polymer material, forming conductive tracks. These carbonized paths progressively reduce surface resistance, exacerbating material degradation. Continuous surface discharges lead to erosion and deeper carbonization, eventually creating conductive pathways that may bridge electrodes, causing complete insulation failure and compromising equipment integrity. Effective management of these phenomena is crucial in maintaining the reliability and safety of high-voltage systems, necessitating careful material selection and environmental control measures.

Thermal management is the most important among the challenges faced in aircraft systems, particularly due to limited convection at low air pressures encountered during wide-body aircraft cruising altitudes, such as 12.2 km, where air pressure is around 18.8 kPa. This reduced convection significantly impacts heat transfer, necessitating careful consideration. Compared to conditions at atmospheric pressure, the maximum permissible current through cables decreases under these circumstances [35, 36]. Moreover, the distribution of electric fields in DC cables is influenced by temperature, which affects conductivity. Understanding the cable's temperature profile is crucial for managing electric field distribution across the insulation. Temperature differentials also influence conductivity gradients within the insulation, contributing to space charge accumulation. Accumulated space charge can alter electric field distributions, potentially leading to dielectric failure and degradation [37]. Polymeric insulation materials are particularly susceptible, as they tend to accumulate space charge more readily beyond their electric field thresholds. To ensure the safe and reliable long-term operation of DC cables in aircraft, it's essential to maintain electric fields below critical thresholds. This necessitates meticulous temperature management throughout the cable's length to design for high safety margins and dependable service life in airborne applications.

3.2 Multilayer multifunctional electrical insulation (MMEI) system

The design of aircraft cables has addressed the above-mentioned challenges by creating MMEI systems. When developing these insulation systems, the key focus is on selecting dielectric materials that offer both high thermal conductivity and dielectric strength. Plastics and polymers, such as aromatic polyimides like Kapton®, are prone to arc tracking due to carbonization under electrical stress. In contrast, aliphatic fluoropolymers exhibit greater resistance to carbon deposits from heat degradation, making them more capable of withstanding wet arc tracking. Additionally, it has been

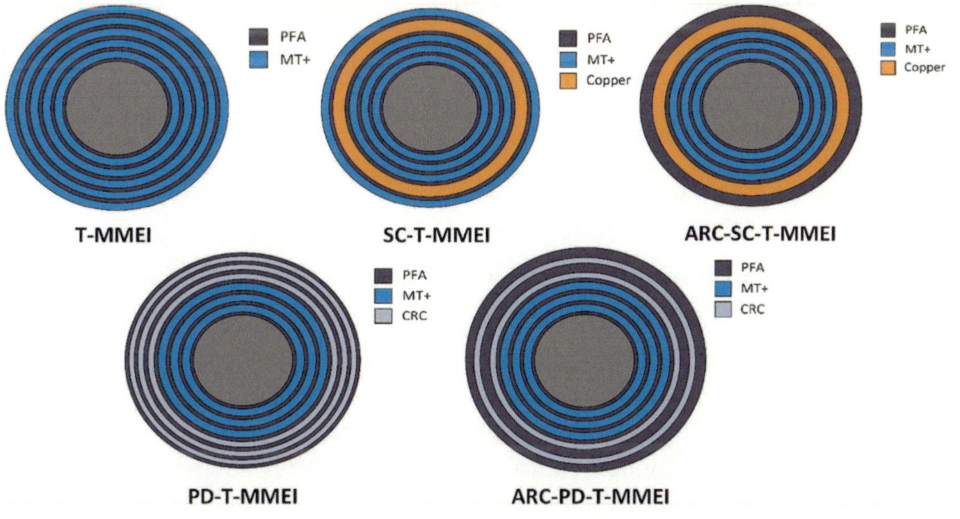

Figure 3.
Different MMEI designs [41].

shown that enhancing polyimide insulation like Kapton® with thin fluoropolymer coatings such as PTFE (Teflon®) can further improve its resistance to wet arc tracking [38].

Research indicates that insulation systems with increased ratios of polyimide demonstrate higher breakdown voltages. Multi-layer configurations incorporating more than three layers of polyimide and fluoropolymer significantly enhance dielectric strength compared to simpler three-layer systems like TKT and Teflon® PFA/Kapton®/Teflon® PFA [39, 40]. In [41], numerous MMEI designs for aviation cables, including T-MMEI, SC-T-MMEI, ARC-SC-MMEI, PD-T-MMEI, and ARC-PD-T-MMEI, were proposed and examined. **Figure 3** shows the different MMEI designs.

These designs utilized varying combinations of polyamides and fluoropolymer films to address specific challenges. Kapton® MT+ was chosen for its superior thermal conductivity and high dielectric strength among polyimide options [42]. However, Kapton® MT+ remains susceptible to arc tracking, mitigated using Teflon® PFA as a protective coating [43]. Additionally, Kapton® CRC, known for its corona-resistant properties [44], was incorporated into some designs to enhance corona-barrier functionality. Here is a concise overview of various MMEI systems used in aviation cables [41]:

- *T-MMEI*: Optimized for thermal performance, featuring a 12-mil wrapped layer comprising Teflon® PFA (0.5 mil film) and Kapton® MT+ (1.5 mil film). Here, 1 mil equals 0.0254 mm.

- *SC-T-MMEI*: Combines thermal optimization and screening with a 5-mil copper layer, 2 mil Teflon® PFA (0.5 mil film) layer, and Kapton® MT+ (1.5 mil film) outer layer.

- *ARC-SC-T-MMEI*: Includes additional arc prevention mechanisms, featuring a 5-mil copper layer and a 4 mil Teflon® PFA jacket.

- *PD-T-MMEI*: Designed for corona-barrier and thermal optimization, with a 4.5 mil wrapped layer of Teflon® PFA (0.5 mil film) and Kapton® CRC (1 mil film) in the outer layer.

- *ARC-PD-T-MMEI*: Combines arc prevention, corona-barrier, and thermal optimization with a 6-mil wrapped layer of Teflon® PFA (2 mil film) and Kapton® CRC (1 mil film).

3.3 Increasing the heat transfer capacity by changing the shape of the cable systems

The shape of a cable significantly affects its maximum current carrying capacity due to variations in radiative and convective heat transfer, both of which depend on the cable's surface area [45–47]. Radiative heat transfer increases with surface area expansion, while convective heat transfer response can vary—increasing, decreasing, or remaining constant—due to numerous influencing factors that may not correlate directly with surface area growth. The radiative heat transfer Q_{ij} between two surfaces i and j can be expressed as:

$$Q_{ij} = \varepsilon_i \sigma A_i F_{ij} \left(T_i^4 - T_j^4 \right) \tag{1}$$

where ε_i is the emissivity of the surface, σ is the Stefan-Boltzmann constant, A_i is the surface area of object i, F_{ij} is the view factor, T_i is the absolute temperature object i, and T_j is the absolute temperature of the object j.

Cuboid and rectangular shapes provide a larger surface area while maintaining the same cross-sectional area and mass compared to cylindrical cables. This increase in surface area enhances the maximum current capacity of a cable due to improved radiative and convective heat transfers. This principle is also applicable to bipolar cable systems. In studies comparing different bipolar cable systems of the same weight and size, such as cylindrical, cuboid, and rectangular shapes, it was found that changing the cable shape to a rectangular form increased the maximum current capacity by approximately 8.3% compared to cylindrical systems and 5.6% compared to cuboid systems [47]. To avoid the issue of theoretically infinite electric fields at the sharp edges of rectangular and cuboid cables, a rounded corner design is implemented. **Figure 4** illustrates the geometry of cylindrical, cuboid, and rectangular cable systems with ARC-SC-T-MMEI insulation systems.

The orientation of a surface significantly influences radiative heat transfer. When bipolar cables are placed near each other, the orientation and distance between them impact radiative heat transfer by affecting the view factor. The view factor determines the portion of radiation emitted by one surface that is received by another and varies with shape, influencing the overall radiative heat transfer between objects. The view factor F_{ij} represents the proportion of radiation emitted from surface i that is intercepted by surface j can be expressed as

$$\mathrm{F_{ij}} = \frac{1}{\mathrm{A_j}} \int_{\mathrm{A_i}} \int_{\mathrm{A_j}} \frac{\cos \theta_\mathrm{i} \cos \theta_\mathrm{j}}{\pi \mathrm{R}^2} \mathrm{dA_i \, dA_j} \tag{2}$$

where, A_i and A_j are the surface area of objects i and j, elemental areas on each surface, dA_i and dA_j, are connected by a line of length R, which forms the polar angles θ_i and θ_j, respectively, with the surface normal n_i and n_j. The analysis of the view factor is crucial in designing bipolar cables. By examining the view factor, engineers or researchers can better understand the radiative heat transfer characteristics between different surfaces of the cable, which is essential for optimizing thermal performance and ensuring the reliability and efficiency of the cable system.

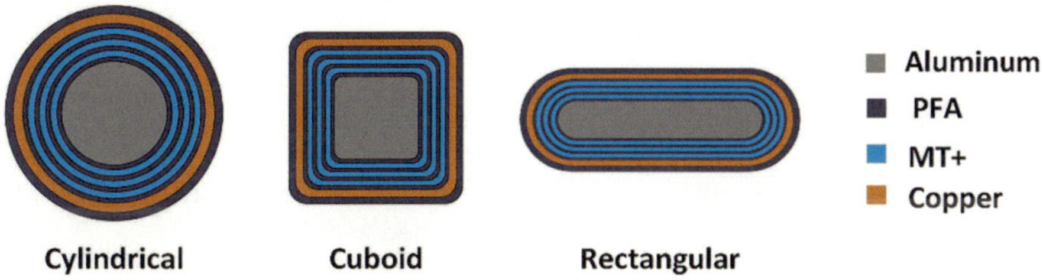

Figure 4.
Geometrical arrangement of three types of cable.

3.4 Coupled electro-thermal model

To attain a high-power-density and low-system-mass EPS for aircraft, cables must support higher currents within existing dimensions or reduce size and mass while maintaining same ampacity. Pursuing the latter, thermal analysis becomes vital in designing cables for optimal heat transfer.

Heat is transferred from the core conductor to the cable surface exclusively through conduction. Within the cable, the heat equation describing this transfer from the inner conductor to the outer jacket can be described as

$$\rho C_p \frac{\partial T}{\partial t} + \nabla.(-k\nabla T) = Q_s \tag{3}$$

where ρ is the density (kg.m^{-3}), k is thermal conductivity (W.(K.m)$^{-1}$), C_P is the specific heat capacity at the constant pressure (J.(kg.K)$^{-1}$), and T is the temperature (K). Q_s stands for the rate of heat generation per unit volume within the material, where it should be transferred from the cable surface via radiative and convective transfers, as can be described as

$$\mathrm{Q_s = Q_r + Q_c} \tag{4}$$

where, Q_r denotes the radiative heat transfer (W.m^{-3}) and Q_c signifies the convective heat transfer (W.m^{-3}). Reducing the cross-sectional area and mass of cables while maintaining a specific ampacity leads to an increase in Q_s due to the increased resistivity of the core conductor resulting from decreased cross-sectional area. Consequently, to counterbalance this rise in total heat loss, adjustments must be made to enhance radiative and convective heat transfers. The heat generated due to Joule losses Q and the relationship between Q_s and Q can be described as follows:

$$Q_s = Q/V \tag{5}$$

$$Q = I^2 R \tag{6}$$

where V represents the volume of the material (m^3), I denotes the conductor current (A) and R denotes the resistance (Ω). The resistance of the conductor varies with temperature according to the following expression for linear resistivity (ρ_e):

$$\rho_e = \rho_0 (1 + \alpha (T_C - T_{ref}) \tag{7}$$

where α is the resistivity temperature coefficient (K^{-1}), T_{ref} represents the reference temperature (K), T_C is the conductor temperature (K), and ρ_0 denotes the electric resistivity (Ω.m) at the reference temperature.

The temperature of the cable is significantly influenced by natural heat convection. The heat equation in the air domain can be described as follows [48]:

$$\rho C_p \frac{\partial T}{\partial t} + \rho C_p u.\nabla T + \nabla.(-k\nabla T) = Q + q + \tau : \nabla u + \frac{-T}{\rho}\frac{\partial \rho}{\partial t}\left(\frac{\partial P}{\partial t} + u.\nabla P\right) \tag{8}$$

where ρ is the air density (kg.m^{-3}), u is the air velocity vector (m.s^{-1}), C_p is the heat capacity at the constant pressure of the air (J.(kg.K)$^{-1}$), k is the thermal conductivity of the air (W.(K.m)$^{-1}$), τ is the viscous tensor (Pa), P is the pressure (Pa), q is the heat flux (W.m^{-3}) and T is the temperature (K). The operator ":" represents the double dot product. The velocity field of the fluid (in this case, air) can be determined by solving the momentum equation and the equation of continuity. These equations are expressed as:

$$\rho\frac{\partial u}{\partial t}+\rho(u.\nabla)u=\nabla.(-PI_I+\tau)+\left(\rho-\rho_{ref}\right)g \tag{9}$$

$$\frac{\partial\rho}{\partial t}+\nabla.(\rho u)=0 \tag{10}$$

where ρ_{ref} is the reference density (kg.m^{-3}), and g is the acceleration of gravity (m.s^{-2}). The electric field distribution can be calculated using

$$E=-\nabla V \tag{11}$$

$$\mathrm{J_e}=\sigma\mathrm{E} \tag{12}$$

where σ is the conductivity (S.m^{-1}), J_e is the current density (A.m^{-2}), and V is the voltage (V). Furthermore, the determination of the steady-state space charge density can be achieved by:

$$\sigma E.\nabla\left(\frac{\varepsilon_e}{\sigma}\right)=\rho_e \tag{13}$$

where ρ_e is the space charge density, and ε_e is the permittivity.

The analysis and discussion in [49, 50] focuses on different types of bipolar cable configurations, including conventional cylindrical, cuboid, rectangular, and coaxial. The study specifically examines the behavior of a $\pm$ 5 kV, 1 kA power cable under a low pressure of 18.8 kPa. **Figure 5** illustrates the 2D geometries of four different bipolar cable systems used for modeling and simulations. These systems employ the previously discussed ARC-SC-T-MMEI configuration for insulation. For simulating heat radiation, the bipolar cable system is enclosed within a duct, represented as the ambient surface in **Figure 5**.

Figure 6 illustrates the weight per unit length and cross-sectional area of the different bipolar cable systems [50]. From the results, it can be shown that the rectangular bipolar cable demonstrates a more efficient weight and cross-sectional area compared to the other three bipolar cable systems. For cylindrical, cuboid, and rectangular bipolar cables, the distance of two inches between the cables results in the lowest weight per unit length and cross-sectional area. At S = 2 inches, rectangular bipolar cables have a weight per unit length of 1.3548 kg/m, making them about 9% and 11.5% lighter than cuboid and cylindrical bipolar cables, respectively, and 26% lighter than coaxial bipolar cables. These findings are consistent with the cross-sectional area measurements, where rectangular bipolar cables at S = 2 inches also have the smallest cross-sectional area compared to the other cable types. This is due to the increased radiative and convective heat transfers of the cables, which result from the greater distance between them. **Figure 7** illustrates the radiative and convective heat fluxes for various types of bipolar cable systems.

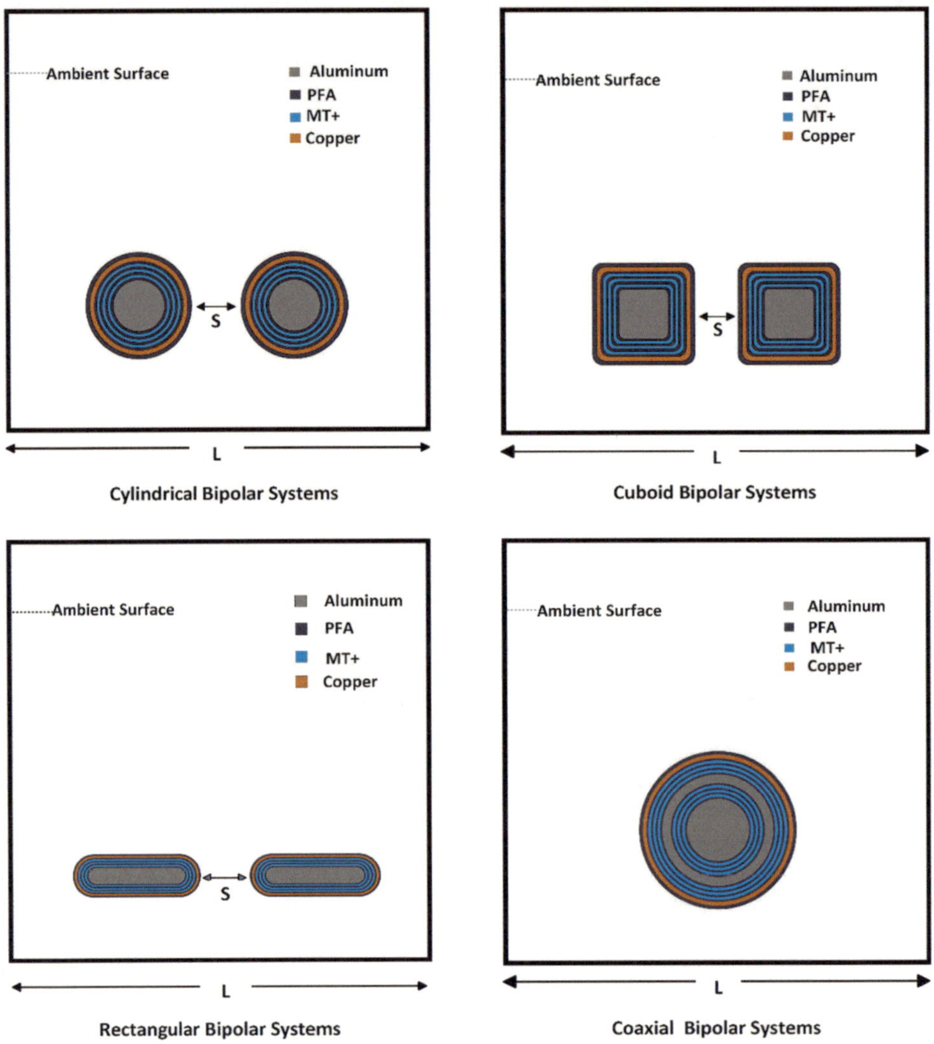

Figure 5.
Geometry of the four types of bipolar cable systems [50].

According to **Figure 7**, the radiative heat flux for cuboid and cylindrical bipolar cables is approximately equal when S = 0 inches, measuring 304 W/m. This value is 9.5% higher than that of coaxial cables, but 21.7% lower than rectangular bipolar cables. As the distance increases, rectangular bipolar cables, despite their smaller cross-sectional area, show increased radiative heat transfer due to a varying view factor. At S = 2 inches, their radiative heat flux is 414.97 W/m, 18.5% and 24.2% higher than cuboid and cylindrical systems, respectively. For convective heat flux, at S = 0 inches, cylindrical, cuboid, and rectangular bipolar cables have lower values due to reduced convection when cables touch. As distance increases, convective heat transfer improves for all systems due to exposure on all sides. Cylindrical and cuboid systems show a slight decrease in convective transfer beyond 0.5 inches, while rectangular cables reach maximum convective transfer at S = 1.5 inches. For a comprehensive analysis, refer to the study mentioned in [50].

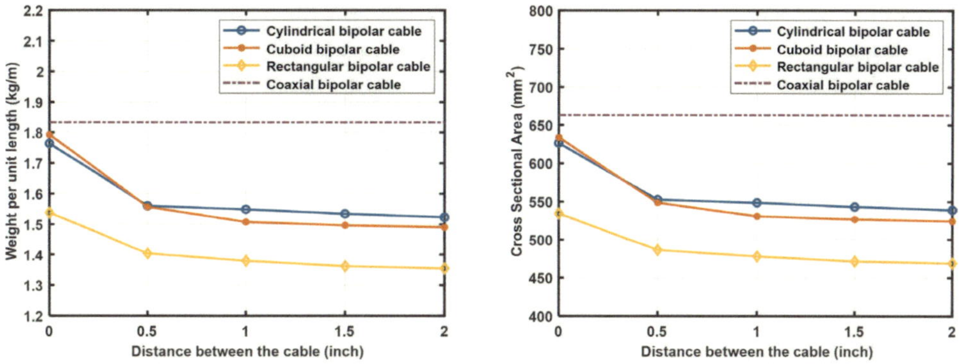

The weight per unit length of four types of bipolar cable systems

The cross-sectional area of four types of bipolar cable systems

Figure 6.
Weight per unit length and cross-sectional area of the four types of cable systems [50].

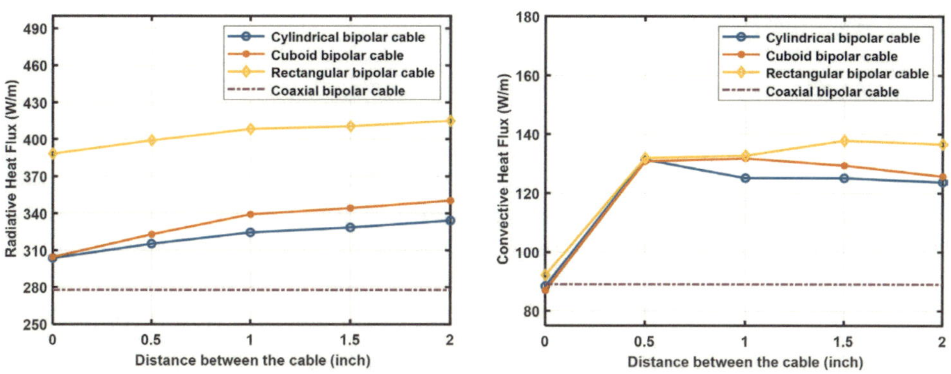

The radiative heat transfer of different bipolar cable systems

The convective heat transfer of different of bipolar cable systems

Figure 7.
Radiative and convective heat fluxes of four types of cable systems [50].

4. Modeling detail in COMSOL multiphysics

4.1 Simulation approach

Power cables are subjected to simultaneous electrical, thermal, and mechanical stresses depending on the applied voltage and current flow. In recent years, the FEM has become crucial for designing reliable insulation systems. FEM simulations enable engineers to analyze insulation materials under various conditions, offering valuable insights into their durability. By accurately modeling complex geometries and material properties, FEM helps create insulation that withstands extreme temperatures, pressures, and environmental challenges, improving overall reliability and performance.

To achieve EPS for aircraft with high power density and low system mass, cable design must focus on either increasing the maximum permissible current without enlarging their cross-sectional area and mass or maintaining the necessary ampacity while reducing the cross-sectional area and mass. In this section, for cable design, the latter approach will be discussed. A comprehensive coupled FEM model is developed using COMSOL Multiphysics software to analyze the temperature and electric field distributions along the MVDC bipolar cable. This model incorporates different modules such as laminar flow, magnetic fields, heat transfer, surface-to-surface radiation, and electric currents. The three modes of heat transfer—radiation, convection, and conduction—are considered during the modeling process.

Three different EPS architectures were introduced and studied in [51] for a wide-body AEA. A maximum ampacity of 1000 A was required for connecting EEUs to busbars. As a result, the bipolar cable was designed and optimized in this work to carry 1000 A of current with the poles' voltages being +5 kV and −5 kV, respectively. The overall size of the cable was optimized for all bipolar cable systems at 1000 A of current to achieve the maximum permissible temperature of 260°C using COMSOL Multiphysics software. After optimization for 260°C, the cross-sectional area and weight per unit length of the cables were evaluated using the Results branch of the COMSOL Multiphysics model tree.

4.2 Cable geometry and simulation settings

Figure 8 illustrates the geometry of the bipolar cylindrical cable systems. To accurately model the three modes of heat transfer under low-pressure conditions, the cables are placed within a duct, which serves as the ambient surface. The duct, made of aluminum alloy, has a square-shaped domain with each side (L) measuring 1 meter in length and a thickness of 1 mm. The emissivity of the aluminum surface varies based on factors such as alloy composition, temperature, oxidation, and surface roughness. For this study, the aluminum alloy is roughly polished, with an emissivity of 0.18, making it suitable as an ambient surface material. The outer surface of the duct is maintained at a constant temperature of 40°C, reflecting typical conditions at the cruising altitude of a wide-body aircraft. The internal pressure inside the duct is considered to be 18.8 kPa. In **Figure 8**, "S" represents the separation between the negative and positive poles, crucial for optimizing the design of future wide-body airplane bipolar cables. Additionally, it is crucial to maintain a sufficient gap between the cables and the sides of the duct to ensure that any potential cable failure does not

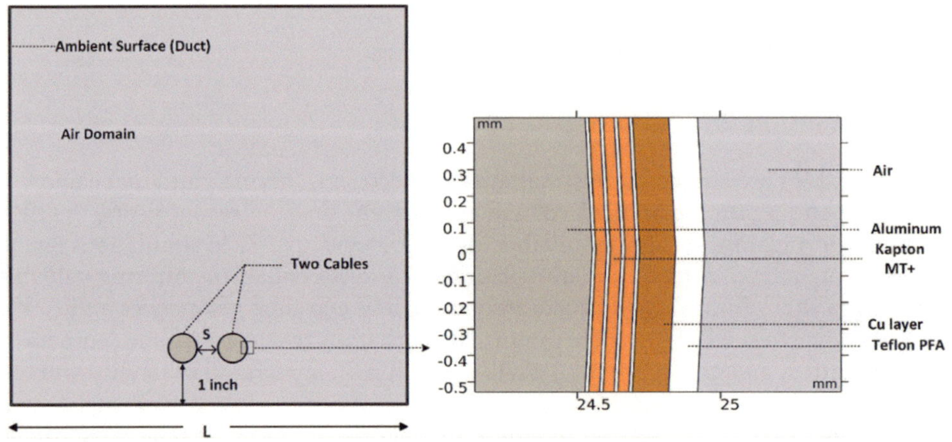

Figure 8.
The geometry of the bipolar cylindrical cable systems in COMSOL.

compromise the safety of the aircraft or its systems. Throughout all simulations conducted in this study, the poles are positioned 1 inch above the duct floor, following the guidelines recommended in [52]. In this bipolar cable system analysis, the previously discussed ARC-SC-T-MMEI insulation system is used, which has a 6.5-mil wrapped layer of Kapton® MT+ and Teflon® PFA. As a screened layer 5 mil copper layer and as a jacket 4 mil Teflon® PFA is used. The zoomed view of this insulation system is shown on the right side of **Figure 8**. The material properties used for this simulation are shown in **Table 1**.

Heat is generated within the core conductor due to joule losses, which then is transferred to the cable surface through conduction. Subsequently, this heat dissipates into the surrounding environment through convection and radiation. To accurately model this process, the study incorporates COMSOL's heat transfer in solids and fluids, laminar flow, magnetic field, electric currents, and surface-to-surface radiation modules. In this analysis, the core conductors are modeled as solid single conductors to simulate heat generation. The magnetic field module in COMSOL is employed and coupled with the heat transfer module to automatically calculate conductor losses. These losses are determined using the electrical resistivity data provided by the magnetic field module. The heat transfer module then incorporates these losses as a heat source when solving the FEM model. This approach ensures a comprehensive simulation of heat transfer dynamics within the cable system, accounting for both electrical and thermal interactions.

The 'coil' node within the Magnetic Field module was utilized in this study, which is normally used to model coils, cables, and other conductors subject to a lumped excitation, such as externally applied current or voltage. Under the coil node as a conductor model, the 'single conductor' model is used. The electric resistivity of the conductor was calculated by the Magnetic Field module using inputs such as current, reference temperature, reference resistivity, and resistivity temperature coefficient. A standard formula available in Section 3, Eq. (7) was employed for this calculation. **Figure 9** shows an illustration of two core conductors designated as heat sources within the Heat Transfer in Solids and Fluids module.

The electric field distribution along the DC cables is influenced by conductivity, which is directly affected by temperature. Concerning the issue of DC conductivity in insulation materials, it is important to consider the non-linear nature of DC conductivity and its implications for electric field distribution. Due to the lack of conductivity data for Kapton® MT+ and Teflon® PFA, Kapton® MT+ was classified as a polyimide material, and Teflon® PFA as an ETFE material. The electrical conductivity of PI and ETFE has been determined by curve-fitting data from [53] using Eq. (14). **Table 2** shows the parameters for calculating the DC conductivity of polymeric materials.

Material	Density ($kg.m^{-3}$)	Dielectric constant	Thermal conductivity ($W.m^{-1}.K^{-1}$)	Dielectric strength ($V.m^{-1}$)
Copper	8960	—	400	—
Kapton® MT+	1420	4.2	0.75	208.5×10^6
Teflon® PFA	2150	2	0.195	256×10^6
Kapton® CRC	1550	3.4	0.2	256×10^6
Aluminum 1350	2705	—	238	—

Table 1.
Materials properties used in analyzed cables [41–44].

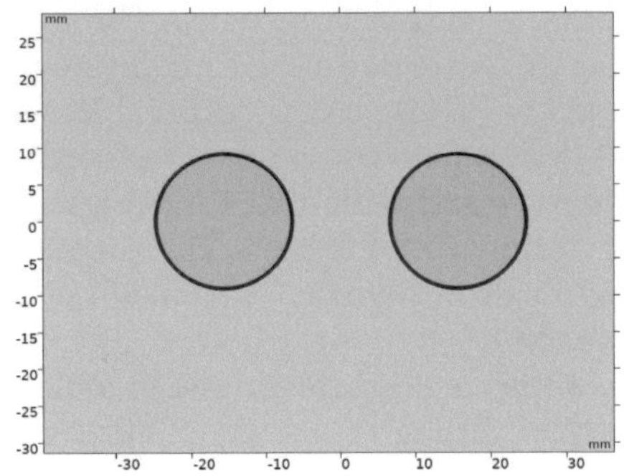

Figure 9.
Two core conductors were chosen as heat sources by the heat transfer in solids and fluids module.

Parameter	*a* (K)	*b* (mm/kV)	σ_0 (S/m)
ETFE	4061	0.03097	2.027e-10
Polyimide	3319	0.05558	1.677e-9

Table 2.
The parameters of the electrical conductivity [50].

$$\sigma(E,T) = \sigma_0 e^{\left(\frac{-a}{T}+bE\right)} \tag{14}$$

where E is the electric field (V.m^{-1}), a is the temperature coefficient, b is the electric field coefficient, and T is the temperature (K).

In screened cables, the screen is grounded. Conversely, for unscreened cables, the outer surface of the cable (the outer boundary of the outermost layer) is grounded. As the ARC-SC-T-MMEI configuration has a 5-mil copper screen layer, those are considered grounds for the simulation. **Figure 10** shows the EC module's ground setting for two cables.

Figure 11 illustrates the meshing arrangement of bipolar cylindrical cables. The precision of computational simulations improves with an increased number of finite elements, as it allows for a more detailed and accurate representation of the model. In this arrangement, a finer mesh is applied to both cables, while a normal mesh is used

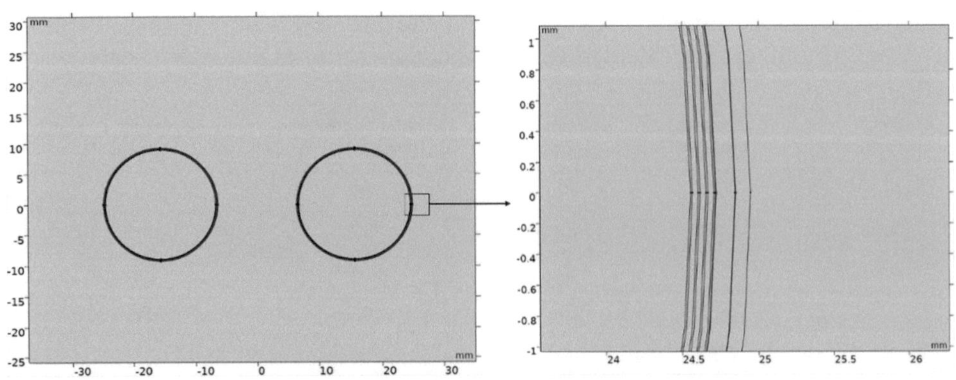

Figure 10.
Ground (0 V) selection of two cables under EC module.

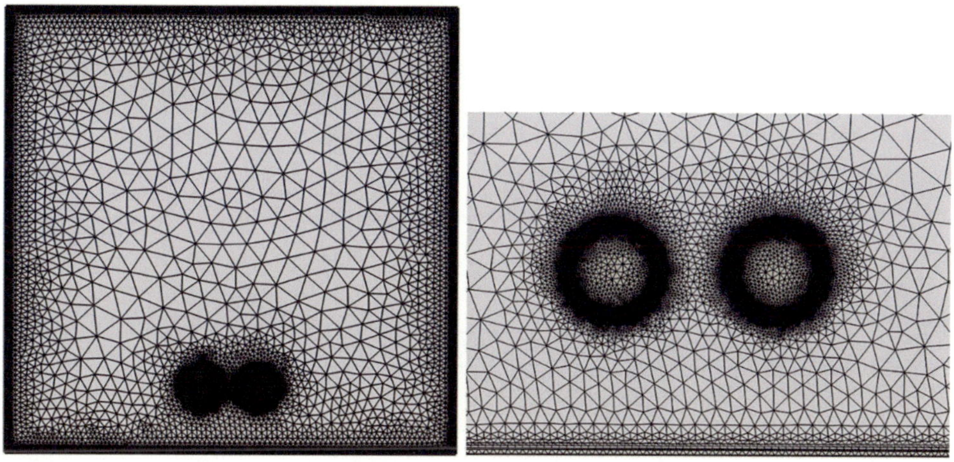

Figure 11.
Meshing arrangements of bipolar cable systems.

for areas like the air and duct sides. The mesh consists of 196,123 vertices and 383,696 elements, including 377,992 triangles, 5704 quadrilaterals, and 59,294 edge elements. This process is managed automatically through adaptive mesh refinement techniques in the software. Adaptive mesh refinement dynamically adjusts the mesh density based on the solution's needs, refining the mesh in regions with high gradients or complex features and coarsening it where less detail is necessary, optimizing both accuracy and computational efficiency.

The simulation incorporates both time-dependent and stationary studies to thoroughly analyze the system. A stationary study is utilized exclusively for the Electric Current module, while time-dependent studies are employed for the other four modules. The simulation runs for 30 hours to ensure it reaches a steady state. **Figure 12** shows the temperature distribution of the bipolar cable systems.

In bipolar cable systems, the overall diameter of the cable is carefully adjusted to ensure it can withstand a maximum permissible temperature of 260°C under a current load of 1000 A. To measure the density, the Surface Integration (∫∫) function located

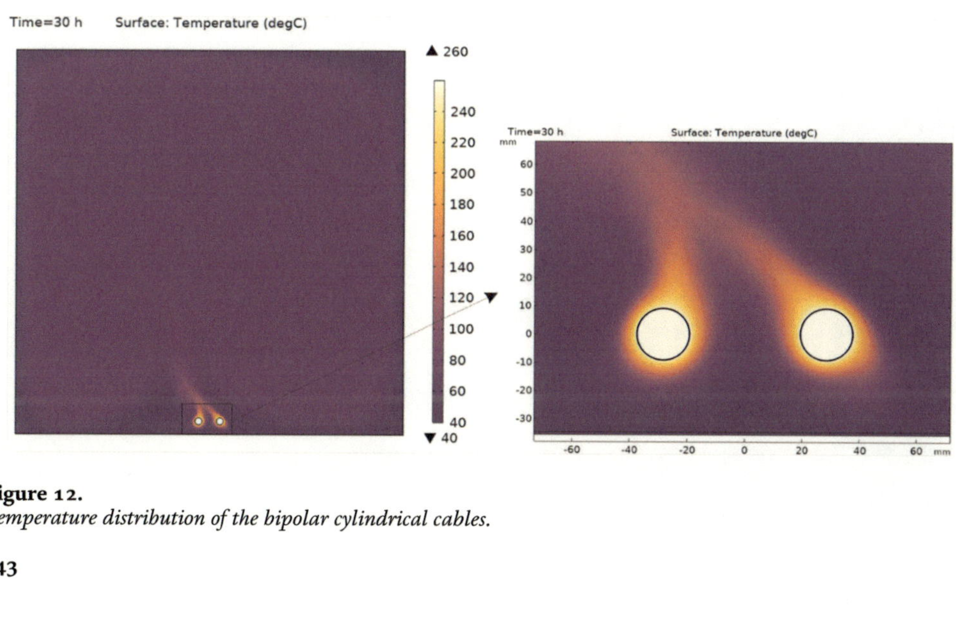

Figure 12.
Temperature distribution of the bipolar cylindrical cables.

Distance between the poles (inch)	Core conductor diameter (mm)	Cable diameter (mm)	Cross-sectional area (mm2)	Weight per unit length (kg.m^{-1})	*J* (g.mm)
0	19.18	19.97	626.25	1.7636	1104.46
0.5	17.97	18.76	552.65	1.5602	862.25
1	17.90	18.69	548.53	1.5488	849.56
1.5	17.81	18.60	543.26	1.5342	833.47
2	17.74	18.53	539.18	1.5320	821.17

Table 3.
Data for optimized cylindrical bipolar cable systems [49].

in the Integration submenu within the Derived Values section was utilized. The derived integration values are used to calculate integrated quantities for each solution in a dataset. The ht.rho function was employed within the expression for surface integration to accurately calculate the density of the chosen domain. For measuring the cross-sectional area, the Surface Measurement function within the Derived Values section was used, which evaluates the surface area over a set of domains in 2D. By using these two functions, the weight per unit length and cross-sectional area of the cables were directly calculated from the COMSOL Multiphysics software. **Table 3** shows the optimized data for cylindrical bipolar cables for different distance between the cables. To simultaneously assess both the weight and dimensions of bipolar cables, a parameter *J* is defined as:

$$J = m.A \tag{15}$$

where *m* is the total cable's mass per unit length (kg.m^{-1}) and *A* is the overall cross-sectional area of the bipolar cable system (m^2).

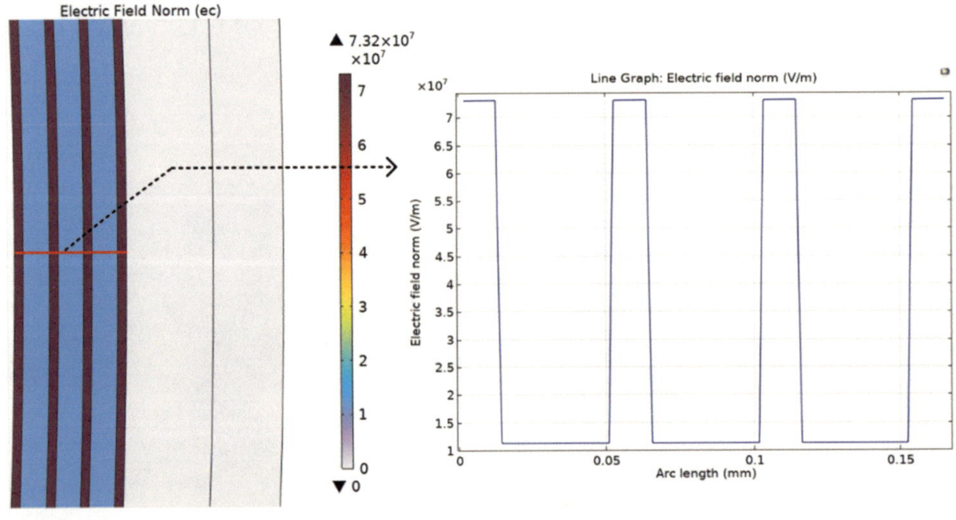

Figure 13.
Electric Field Distribution of bipolar cables.

The Electric Currents (EC) module was employed to analyze the insulation characteristics of the cables. The electric field distribution across the insulation in DC cables results from the combined effects of the current field and Poisson's electric field. The EC module calculates the space charge density within the cables and incorporates the associated Poisson's electric field into the current field. By solving the electrical conductivity expression, the EC module determines the electric field across the insulation, encompassing both the current and Poisson's electric fields within its solution. Thus, the resultant electric field across the designed cables represents a comprehensive total electric field. **Figure 13** shows the electric field distribution of the bipolar cylindrical cables. By using the "Cut Line 2D" feature (red line in **Figure 13**), the electric field distribution across the MMEI insulation is determined.

5. Fabrication and testing

In previous discussions, the simulation and modeling procedures for bipolar MVDC power cables in wide-body AEA were explored. To ensure proper electrical insulation performance of these power cables, experimental testing, in addition to modeling and simulations, is required. The dielectric strength and time-to-failure of these designs must be validated through experimental evaluation. Specifically, comprehensive testing of power cables with MMEI configurations is essential for assessing their commercial viability. Consequently, samples of the designed MMEI structures need to be fabricated for testing.

The prototyping process of MMEI insulation, both as flat samples and MVDC power cables with MMEI configurations, is discussed here. The necessity of optimizing manufacturing conditions to ensure uniformity and eliminate defects is emphasized, aiming to produce high-quality samples for testing.

5.1 Compression molding process

The compression molding process is a crucial method for creating MMEI samples. Several factors, such as mold size, compression pressure, duration, and temperature, significantly influence this process. Careful attention to these techniques and optimization of these parameters is essential for enhancing the quality of the fabricated samples. The process can be outlined as follows:

1. *Design and Fabrication of Molds*: The initial step involves designing molds that match the specifications of the MMEI configurations. The mold size must accommodate the desired dimensions of the dielectric layers and the overall cable structure. For fabricating flat samples, high-precision A2 steel molds with lengths of 6 inches and widths of 1, 2, and 6 inches are used [54].

2. *Layer Arrangement*: The dielectric layers must be precisely arranged in the mold to ensure uniform distribution and proper adhesion during the molding process. The formation of voids depends on the arrangement of different polymeric layers. It has been shown that spiraling different dielectric layers in opposite directions improves the PD performance of the fabricated cable samples [55].

3. Optimization of Compression Molding Parameters:

- *Compression Pressure*: Sufficient pressure must be applied during molding to bond the layers without introducing voids. For fabricating flat samples, high-temperature C3 Connor sealing clips are used. The quantity of clips for each mold size is modified to ensure uniform compression pressure and to prevent void formation in the manufactured MMEI system [54]. For fabricating MMEI cable samples, adjustable T-bolt hose clamps are used [55–57].

- *Duration*: The molding duration should be sufficient for proper bonding of polymeric materials without causing degradation.

- *Temperature*: The molding temperature must be optimized to ensure material flexibility and effective bonding without compromising dielectric properties. PD tests on different flat samples have been conducted to optimize the duration and temperature, with the best results achieved at 360°C and 40 minutes [54].

4. *Quality Control*: Post-molding, the samples should undergo rigorous quality control checks to identify and rectify any defects. This includes visual inspections, dielectric testing, and mechanical assessments.

By meticulously managing these steps and optimizing the compression molding parameters, high-quality MMEI samples can be produced for experimental testing. **Figure 14** shows the fabricated flat MMEI samples and MMEI cable samples. This approach is crucial for validating the performance of MVDC power cables in AEA applications and ensuring their commercial viability.

5.2 PD test setup and experimental procedure

The experimental setup for analyzing PD is depicted in **Figure 15**. The high-voltage DC source utilized in this study is the PHENIX Model 4100–10, which can generate an output voltage of up to 100 kV DC and measure leakage currents ranging from 0.01 to 20,000 μA DC. The OMICRON MPD 800, a state-of-the-art system for

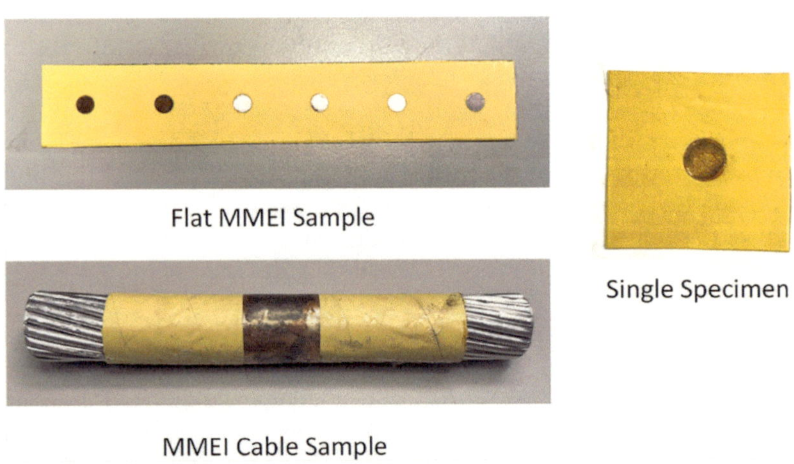

Figure 14.
Fabricated samples for testing.

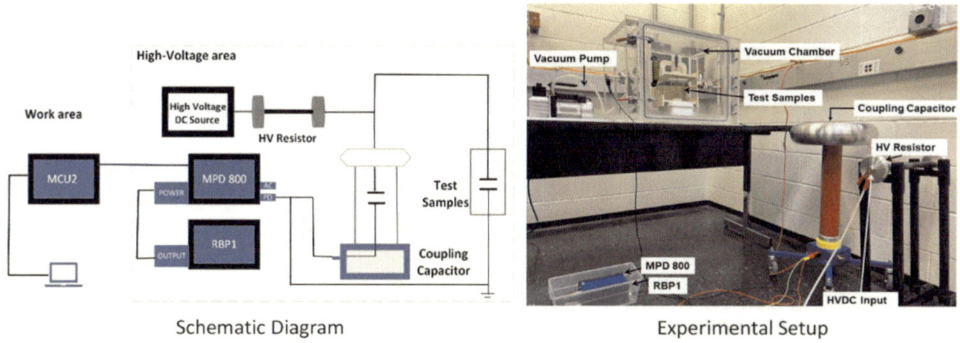

Figure 15.
PD test setup.

PD measurement and analysis, was employed. To detect PD signals, a coupling capacitor was used per the standards specified in IEC 60270 [58]. An HV resistor capable of withstanding voltages up to 100 kV was also employed to control the high current directed into the OMICRON MPD 800 device. As an electrode setup, a sphere-sphere electrode was used for measuring the PD signal [54]. This setup can be used as a reference for doing the PD test for the MVDC system.

6. Conclusion

Transitioning to AEA is a crucial step towards achieving net-zero GHG emissions in the transportation sector, particularly within aviation. The power cable, as a fundamental component of an aircraft's EPS, plays a significant role in achieving a low system mass EPS. Given the limited heat convection at the cruising altitudes of wide-body AEA, thermal management of the power cable is one of the most critical challenges. The introduction of MMEI systems represents a significant advancement over traditional single-layer insulation, effectively addressing these thermal challenges. This book chapter provides a comprehensive overview of the design, fabrication, and testing processes for aircraft MVDC power cables, supported by detailed coupled electrical, thermal, and CFD models. These models, implemented in COMSOL, enable optimization of cable geometries and provide valuable insights and methodologies for developing efficient and reliable EPSs in AEA. This work serves as an essential reference for researchers and engineers and as an educational resource for understanding the complexities of power cable design in the context of electrified aviation.

Acknowledgements

This work was supported in part by the Advanced Research Projects Agency-Energy (ARPA-E), U.S. Department of Energy, under Award DE-AR0001677.

Conflict of interest

The authors declare no conflict of interest.

Author details

Anoy Saha and Mona Ghassemi*
Zero Emission, Realization of Optimized Energy Systems (ZEROES) Laboratory, The University of Texas at Dallas, Richardson, TX, USA

*Address all correspondence to: mona.ghassemi@utdallas.edu

References

[1] United Nations. Department of Economic and Social Affairs [Internet]. Available from: https://sdgs.un.org/goals/goal13 [Accessed: June 08, 2024]

[2] US EPA. Sources of Greenhouse Gas Emissions [Internet]. 2024. Available from: https://www.epa.gov/ghgemissions/sources-greenhouse-gas-emissions [Accessed: June 08, 2024]

[3] IEA. Aviation [Internet]. Available from: https://www.iea.org/energy-system/transport/aviation [Accessed: June 08, 2024]

[4] Alternative Fuels Data Center. Electric Vehicle Benefits and Considerations [Internet]. Available from: https://afdc.energy.gov/fuels/electricity-benefits [Accessed: June 08, 2024]

[5] Kumar M, Panda KP, Naayagi RT, Thakur R, Panda G. Comprehensive review of electric vehicle technology and its impacts: Detailed investigation of charging infrastructure, power management, and control techniques. Applied Sciences. 2023;**13**(15):8919. DOI: 10.3390/app13158919

[6] Ansell PJ, Haran KS. Electrified airplanes: A path to zero-emission air travel. IEEE Electrification Magazine. 2020;**8**:18-26. DOI: 1109/mele.2020.2985482

[7] United States. 2021 Aviation Climate Action Plan [Internet]. 2021. Available from: https://www.faa.gov/sites/faa.gov/files/2021-11/Aviation_Climate_Action_Plan.pdf [Accessed: June 08, 2024]

[8] Adu-Gyamfi BA, Good C. Electric aviation: A review of concepts and enabling technologies. Transportation Engineering. 2022;**9**:100134. DOI: 1016/j.treng.2022.100134

[9] IATA. Our Commitment to Fly Net Zero by 2050 [Internet]. Available from: https://www.iata.org/en/programs/environment/flynetzero/ [Accessed: June 12, 2024]

[10] Schefer H, Fauth L, Kopp TH, Mallwitz R, Friebe J, Kurrat M. Discussion on electric power supply systems for all electric aircraft. IEEE Access. 2020;**8**:84188-84216. DOI: 10.1109/access.2020.2991804

[11] Gemin P, Kupiszewski T, Radun A, Pan Y, Lai R, Zhang D, et al. Architecture, voltage, and components for a turboelectric distributed propulsion electric grid (AVC-TeDP) [Internet]. 2015. Available from: https://ntrs.nasa.gov/api/citations/20150014583/downloads/20150014583.pdf [Accessed: June 08, 2024]

[12] Madonna V, Giangrande P, Galea M. Electrical power generation in aircraft: Review, challenges, and opportunities. IEEE Transactions on Transportation Electrification. 2018;**4**:646-659. DOI: 1109/tte.2018.2834142

[13] IEC 60287-1-1:2006. Electric Cables-Calculation of the Current Rating-Part 1–1: Current Rating Equations (100% Load Factor) and Calculation of Losses-General. Geneva, Switzerland: IEC; 2006. pp. 1-136

[14] IEC TR 62095:2003. Electric Cables-Calculations for Current Ratings-Finite Element Method. Geneva, Switzerland: IEC; 2003. pp. 1-69

[15] Hwang C-C, Jiang Y-H. Extensions to the finite element method for thermal

analysis of underground cable systems. Electric Power Systems Research. 2003; **64**:159-164. DOI: 10.1016/s0378-7796(02)00192-x

[16] Christou I, Nelms A, Husband M, Cotton I. Choice of optimal voltage for more electric aircraft wiring systems. IET Electrical Systems in Transportation. 2011;**1**:24-30. DOI: 10.1049/iet-est.2010.0021

[17] Maldonado MA, Korba GJ. Power management and distribution system for a more-electric aircraft (MADMEL). IEEE Aerospace and Electronic Systems Magazine. 1999;**14**:3-8. DOI: 10.1109/62.811084

[18] Homeyer WG, Bowles EE, Lupan SP, Walia PS, Maldonado MA. Advanced power converters for more electric aircraft applications. In: Proceedings of the Thirty-Second Intersociety Energy Conversion Engineering Conference (IECEC-97). HI, USA: Honolulu; 1997. pp. 591-596

[19] Sarlioglu B, Morris CT. More electric aircraft: Review, challenges, and opportunities for commercial transport aircraft. IEEE Transactions on Transportation Electrification. 2015;**1**: 54-64. DOI: 10.1109/tte.2015.2426499

[20] Chen J, Wang C, Chen J. Investigation on the selection of electric power system architecture for future more electric aircraft. IEEE Transactions on Transportation Electrification. 2018; **4**:563-576. DOI: 10.1109/tte.2018.2792332

[21] Alcorta-Garcia E, Zolghadri A, Goupil P. A nonlinear observer-based strategy for aircraft oscillatory failure detection: A380 case study. IEEE Transactions on Aerospace and Electronic Systems. 2011;**47**:2792-2806. DOI: 10.1109/taes.2011.6034665

[22] Steinbauer T, Leiprecht G, Havraneck W. High performance HIL real time for the Airbus A380 electrical backup hydraulic actuators. In: 2004 Mini Symposia UKACC Control. UK: Bath; 2004. pp. 71-74

[23] No-Bleed Systems. Saving Fuel and enhancing operational efficiencies [Internet]. Available from: http://www.boeing.ch/commercial/aeromagazine/articles/qtr_4_07/AERO_Q407_article2.pdf [Accessed: June 08, 2024]

[24] Xu K, Xie N, Wang C, Wang Y. A comprehensive simulation model and stability analysis for power system of more electrical aircraft. In: 2016 IEEE International Conference on Aircraft Utility Systems (AUS). Beijing, China: IEEE; 2016. pp. 219-226

[25] Wheeler P. Technology for the more and all electric aircraft of the future. In: In: 2016 IEEE International Conference on Automatica (ICA-ACCA). Curico, Chile: IEEE; 2016. pp. 1-5

[26] NASA. Glenn Research Center. Subsonic Single Aft Engine (SUSAN) Aircraft [Internet]. 2024. Available from: https://www1.grc.nasa.gov/aeronautics/eap/airplane-concepts/susan/ [Accessed: June 08, 2024]

[27] NASA. Glenn Research Center. Single-Aisle Turboelectric Aircraft with Aft Boundary Layer Propulsion (STARC-ABL) [Internet]. 2023. Available from: https://www1.grc.nasa.gov/aeronautics/eap/airplane-concepts/starc-abl/ [Accessed: June 08, 2024]

[28] NASA. Glenn Research Center. N3-X [Internet]. 2023. Available from: https://www1.grc.nasa.gov/aeronautics/eap/airplane-concepts/n3x/ [Accessed: June 08, 2024]

[29] ESAero – Empirical Systems Aerospace. ECO-150 [Internet]. 2024. Available from: https://www.esaero.com/core-technology-innovations/eco-150 [Accessed: June 08, 2024]

[30] Dricot F, Reher HJ. Survey of arc tracking on aerospace cables and wires. IEEE Transactions on Dielectrics and Electrical Insulation. 1994;**1**:896-903. DOI: 10.1109/94.326657

[31] Muhamedin FL, Piah M, a. M, Othman NA. Modelling on tracking test condition of polymer nanocomposite using finite element simulation. Telecommunication Computing Electronics and Control (Telkomnika). 2015;**13**:1194. DOI: 10.12928/telkomnika.v13i4.2362

[32] Rui R, Cotton I. Impact of low pressure aerospace environment on machine winding insulation. In: IEEE International Symposium on Electrical Insulation. San Diego, CA, USA: IEEE; 2010. pp. 1-5

[33] Esfahani AN, Shahabi S, Stone G, Kordi B. Investigation of corona partial discharge characteristics under variable frequency and air pressure. In: 2018 IEEE Electrical Insulation Conference (EIC). San Antonio, TX, USA: IEEE; 2018. pp. 31-34

[34] Cotton I, Nelms A. Higher voltage aircraft power systems. IEEE Aerospace and Electronic Systems Magazine. 2008; **23**:25-32. DOI: 10.1109/maes.2008.4460728

[35] Azizi A, Ghassemi M, Lehr JM. Heat transfer challenges for MVDC power cables used in wide body all electric aircraft under low pressures. IEEE Access. 2022;**10**: 111811-111819. DOI: 10.1109/access.2022.3216300

[36] Azizi A, Ghassemi M. An analysis of parameters affecting ampacity in aircraft bipolar MVDC power cables via coupled electrical, thermal, and computational fluid dynamic modelling. IET High Voltage. 2024; To be published. DOI: 10.1049/hve2.12452

[37] Fabiani D, Montanari GC, Laurent C, Teyssedre G, Morshuis PHF, Bodega R, et al. HVDC cable design and space charge accumulation. Part 3: Effect of temperature gradient [feature article]. IEEE Electrical Insulation Magazine [Internet]. 2008;**24**:5-14. DOI: 10.1109/mei.2008.4473049

[38] Cahill PL, Dailey JH. Aircraft Electrical Wet-Wire Arc Tracking [Internet]. Astrophysics Data System (ADS). 1988. Available from: https://ui.adsabs.harvard.edu/abs/1988faa..rept.....C/abstract [Accessed: June 08, 2024]

[39] Shin EE, Scheiman DA, Lizcano M. Lightweight, durable, and multifunctional electrical insulation material systems for high voltage applications. In: AIAA/IEEE Electric Aircraft Technologies Symposium (EATS). Cincinnati, OH, USA: IEEE; 2018. pp. 1-21

[40] Shin EE. Development of high voltage micro-multilayer multifunctional electrical insulation (MMEI) system. In: AIAA/IEEE Electric Aircraft Technologies Symposium (EATS). Indianapolis, IN, USA: IEEE; 2019. pp. 1-14

[41] Azizi A, Ghassemi M. Design of high power density MVDC cables for wide-body all electric aircraft. IEEE Transactions on Dielectrics and Electrical Insulation. 2023;**30**: 2315-2324. DOI: 10.1109/tdei.2023.3285849

[42] DuPont™ Kapton® MT+ [Internet]. Available from: https://materials-direct.com/wp-content/uploads/2021/05/K-MT-Eng-2019.pdf [Accessed: June 08, 2024]

[43] DuPont™ Teflon® PFA [Internet]. Available from: https://catalog.cshyde.com/Asset/Data%20Sheet%2023-__PFA%20DuPont%20PFA%20Film%20Data%20Sheet.pdf [Accessed: June 08, 2024]

[44] DuPont™ Kapton® CRC [Internet]. Available from: https://www.dupont.com/content/dam/dupont/amer/us/en/ei-transformation/public/documents/en/EI-10175-Kapton-100CRC-Data-Sheet.pdf [Accessed: June 08, 2024]

[45] Azizi A, Saha A, Ghassemi M. A cuboid geometry design for MVDC power cables for using in future all electric wide body aircraft. In: IEEE Conference on Electrical Insulation and Dielectric Phenomena (CEIDP). East Rutherford, NJ, USA: IEEE; 2023. pp. 1-4

[46] Saha A, Azizi A, Ghassemi M. An optimal bipolar MVDC coaxial power cable design for envisaged all electric wide body aircraft. In: IEEE Conference on Electrical Insulation and Dielectric Phenomena (CEIDP). East Rutherford, NJ, USA: IEEE; 2023. pp. 1-4

[47] Saha A, Ghassemi M. MVDC bipolar power cables with rectangular geometry design for envisaged all-electric wide-body aircraft. In: IEEE Texas Power and Energy Conference (TPEC). College Station, TX, USA: IEEE; 2024. pp. 1-5

[48] Panton RL. Incompressible Flow. 2nd ed. Hoboken, New Jersey, USA: John Wiley and Sons; 1996. DOI: 10.1002/9781118713075

[49] Saha A, Azizi A, Ghassemi M. Optimal bipolar MVDC power cable designs for future wide-body all electric aircraft. IEEE Transactions on Dielectrics and Electrical Insulation. 2024. DOI: 10.1109/tdei.2024.3355033

[50] Saha A, Ghassemi M. Optimized design of ±5 kV, 1 kA rectangular power cable for a low pressure of 18.8 kPa for envisioned all-electric wide-body aircraft. IEEE Access. 2024;**12**:28654-28665. DOI: 10.1109/access.2024.3368437

[51] Ghassemi M, Barzkar A, Saghafi M. All-electric NASA N3-X aircraft electric power systems. IEEE Transactions on Transportation Electrification. 2022;**8**: 4091-4104. DOI: 10.1109/tte.2022.3158186

[52] AC 25.1701-1 Certification of Electrical Wiring Interconnection Systems on Transport Category Airplanes [Internet]. 2024. Available from: https://www.faa.gov/regulations_policies/advisory_circulars/index.cfm/go/document.information/documentID/73476 [Accessed: June 08, 2024]

[53] Baferani MA, Li C, Shahsavarian T, Ronzello J, Cao Y. High temperature insulation materials for DC cable insulation — Part I: Space charge and conduction. IEEE Transactions on Dielectrics and Electrical Insulation. 2021;**28**:223-230. DOI: 10.1109/tdei.2020.009167

[54] Chowdhury S, Saha A, Rahman MA, Kalakonda SP, Ghassemi M. Study of partial discharge characteristics of optimally fabricated MMEI flat samples under DC and their DC dielectric strength for applications in envisaged all-electric wide-body aircraft. IEEE Access. Accepted for publication. 2024; **12**:88547-88557. DOI: 10.1109/ACCESS.2024.3418789

[55] Chowdhury S, Azizi A, Saha A, Rahman MA, Ghassemi M, Lehr J. An optimal approach to fabricate MVDC multilayer insulation systems as flat samples for wide-body all-electric aircraft. In: IEEE Dallas Circuit and Systems Conference (DCAS). Richardson, Texas, USA: IEEE; 2024

[56] Rahman MA, Saha A, Ghassemi M. Optimized fabrication process and PD characteristics of MVDC multilayer insulation cable systems for next generation wide-body all-electric aircraft. Energies. 2024;**17**(12):3040. DOI: 103390/en17123040

[57] Saha A, Rahman MA, Chowdhury S, Ghassemi M, Lehr J. Prototyping aircraft MVDC power cables with optimal multilayer multifunctional electrical insulation systems. In: IEEE International Power Modulator and High Voltage Conference (IPMHVC). Indianapolis, Indiana, USA: IEEE; 2024

[58] IEC 60270:2000. High-voltage test techniques—Partial discharge measurements. In: International Electrotechnical Commission. Geneva, Switzerland: IEC; 2000